I0049236

First published in 2025 by Headfirst Publishing
Based in Brisbane Australia

Typeset in Georgia Regular

National Library of Australia Cataloguing-in-Publication data:

Creator	Tony J. Ryan, author
Title	How To Predict The Future... most of the time: Merging AI, human intuition, and predictive analysis to forecast what's up ahead
ISBN	978-0-9577267-2-7 (pbk) 978-0-9577267-3-4 (ebook)
Notes	Includes index
Subjects	Future, The Social prediction Self-help

Typesetting by Annette Vandermaat (Graphic Design by AV)
Cover design by Imran Khaliq (Pixel-Lab)

Disclaimer
The material in this publication is of the nature of general comment only, and does not represent professional or financial advice. It is not intended to provide specific guidance for particular circumstances and it should not be relied upon as the basis for any decision to take action or not to take action on any manner which it covers. Readers should obtain professional advice where appropriate before making any such decision. To the maximum extent permitted by law, the author and publisher disclaim all responsibility and liability to any person, arising directly or indirectly from any person taking or not taking action based on the information in this publication.

Acknowledgments

While predicting the future is crucial, the past and present remain profoundly significant. I am deeply thankful to those who have shaped this book with their invaluable contributions.

My heartfelt thanks to Joan Dalton, Alie Blackwell, John Hale, and Catherine McCullough for their insightful perspectives, which greatly enhanced this work. I also extend my gratitude to Terry O'Neill for his advice in the chapter on 'Life's a Gamble'.

Special thanks to my incredible family for their unwavering support, and to my wonderful wife, Lana, whose love knows no bounds.

Additionally, I appreciate the many educators and professionals from Australia, New Zealand, Singapore, Hong Kong, China, Malaysia, the United Kingdom, Mexico, the United States, and Canada who have engaged with me and enriched this journey.

This book is dedicated to every one who passionately believes and predicts that our future can be human-centred and extraordinary.

THE THREE CORE QUESTIONS

This book addresses three specific questions:

Q. How can you really predict the future?

Q. Can your future be altered?

Q. How might prediction help you to live a more fulfilling life?

The book content will provide insights into these questions as you progress. If you prefer to find answers sooner, you can refer to the end of the book. However, to fully enhance your predictive understanding, I recommend exploring the concepts and practical ideas presented throughout the book before reaching the conclusion.

ABOUT THE AUTHOR

Tony Ryan has supported over 1,000 businesses and educational organizations across 12 countries on future pathways, and has keynoted at more than 100 conferences. He is the author of books on innovative thinking and lifelong learning, including *Thinkers Keys*, *The Ripple Effect*, and his latest release, *The Next Generation*, published by Wiley & Sons. Ryan's promotion for this book included appearances on national TV and a multitude of radio and newspaper interviews.

You can access ongoing updates on this book and its core topic at *predictfutures.au*

This will include a specific section for Book Updates.

HOW TO
PREDICT
THE
FUTURE
MOST OF THE TIME

MERGING AI, HUMAN INTUITION, AND PREDICTIVE
ANALYSIS TO FORECAST WHAT'S UP AHEAD

TONY RYAN

PREFACE

Prediction is very difficult, especially if it's about the future.
NIELS BOHR

Can We Really Predict the Future?

Imagine having the ability to predict the future. In our turbulent world, the physical, social, mental, and financial benefits would be transformative. But how can you foresee what hasn't happened yet? Is prediction merely an educated guess tinged with luck, or can you forecast future events with a high degree of probability? The answer to the latter question is a carefully qualified "most of the time". This book delves into why, how, and when you can predict most of your future, most of the time.

The Limits and Possibilities of Prediction

While you can't predict everything—like winning lottery numbers, sudden accidents, or rapid stock market crashes—many events can be forecasted using predictive analysis, probability calculations, and human insight. Complex geopolitical events may elude precise prediction due to their myriad factors, but other areas hold promise. This book will explore how you can apply these techniques. Unlike many accomplished futurists who offer insightful narratives, my approach is to equip you with the tools to predict (most of) your future.

The Value of Predictive Capacity

For those striving for extraordinary achievement, this book is particularly relevant. High achievers often engage in prediction. Whether it's an outstanding sportsperson anticipating an opponent's next move, an exceptional teacher or parent gauging the long-term impact of quality guidance for children, or an astute businessperson forecasting industry trends, predictive capacity adds significant value to our lives.

Building Predictive Capacity

This book will cover the basics of probability and how to predict by observing natural cycles and patterns. We will scrutinize ancient and modern superstitions and strategies for their scientific credibility and explore the potential financial benefits of predictive analysis. Topics such as DNA testing, AI, trendspotting, science fiction, and genetic influence will be assessed. Everyday techniques like horoscopes and crystal ball readings will be questioned, while unusual abilities such as intuitive prediction will be assessed for their credence.

Historical Context and Human Development

Prediction has always driven human development. Leaders historically gained power and prestige by demonstrating—authentically or otherwise—an ability to predict the future. Chapter 1 will explore a fascinating array of predictive tools from the past, from palm readings to star mapping, illustrating the enduring value of predictive capability. For everyday people, foretelling events like the arrival of rains could mean the difference between life and death.

Let's embark on this exploration of prediction by learning from the past, aiming to harness these insights for a better future.

CHAPTER 1: PREDICTING FROM THE PAST

Knowledge is telling the past.
Wisdom is predicting the future.

W. TIMOTHY GARVEY

Living With Superstition

Superstitious practices dominated early civilizations. The ancient Romans believed that you could cure hiccups and a runny nose by kissing a female mule in the nostrils. History does not record how the mule felt about this practice. Commonfolk in the Middle Ages would say "bless you" after someone sneezed. According to their beliefs, a sneeze gave Satan the chance to enter your body, and the person who sneezed needed God's help to exorcise the devil.

In modern life, we still believe in many of these superstitions (except perhaps for the female mule). We might think we have developed into an advanced society, and yet many of us still avoid walking under a ladder, or crossing the path of a black cat, or refusing to have 666 on our vehicle license plates. We steadfastly maintain faith in our special Lotto numbers, we are unsettled by a 13th floor in hotel buildings, and we shudder at the thought of breaking a mirror. [1]

Many of these superstitions are hundreds of years old, and were created because people hoped to influence and even predict their own future. [2] In earlier times, lives were compromised by a fear of the unknown world up ahead. To counteract this, people engaged in specific activities such as wearing a special amulet or crossing themselves three times, in the hope that good luck would befall them. It was their version of creating a better world for themselves in their future.

In the Middle Ages, partly because of widespread illiteracy, superstitions would spread rapidly through constant gossip. It was similar to today's social media. Many people back then were convinced that demons and evil spirits could easily enter their hut and take away their souls. Village elders, and anyone who craved power, took advantage of these superstitions. To introduce a new superstition, or to maintain an old one, encouraged the commonfolk to pay respect to you. Witchdoctors, soothsayers, prophets, visionaries, shamans, and oracles feature all through human history, and many of them based their control on fear that was generated by superstitions. Today we attach our need to be saved on so-called gurus, famous authors and influencers.

In some cultures, the pointing of the bone by the shaman in a person's direction could lead to his or her death. Perhaps a life might be determined by the movement of flames in a fireplace, or by the toss of a coin. Imagine watching the coin spinning, and knowing that if it ended up on Heads, you would lose yours. Police forces were not common until well into the 1800s in most parts of the world, and many people had to fend for themselves, relying predominantly on the goodwill of their fellow citizens. Superstitions became part of the ground rules for organising present and future life.

Behaviours based upon superstitions help you to believe that you can predict your own fate. It can be uncomfortable not knowing what will happen up ahead, and so you might play along with the superstition to feel more positive about the future. Almost any action can be your personal superstition. If you wore a specific set of socks the last time you played well at your sport, you might wear them again the next time, and sometimes without washing them.[3] What superstitions do you sometimes practise?

Prediction Strategies from Earlier Times

Superstitious practices were not the only way that our ancestors tried to influence or predict the future. Crystal ball readings, palmistry, numerology, Tarot cards, water divining, the reading of tea leaves, and astrological horoscopes featured strongly in past ages.[3] Some still do today. Many of these practices ended their title with the suffix 'mancy', which refers to the practice of reading the future.

They included anthropomancy (analyse human entrails), bibliomancy (open a book and predict something with the first word you see), dactylomancy (suspend a wedding ring with a fine thread over a table with letters and allow the swinging movement to spell out your answer), pyromancy (predict the future through the movement of flames) and oneiromancy (predict the future with dreams).[4]

An endless range of other strange prediction options have thankfully passed their use-by date. Around the 1850s in Northern Ireland, women forecast the character of their future husbands by analysing the slimier parts of a herring. I'm not sure what that says about the quality of the men in that time. *The Old Egyptian Fortune-Teller's Last Legacy* was published in London in 1775 and claimed that one way to foresee the future was by interpreting the moles on your face and body. For example, "a mole on the buttock denotes honour to a man and riches to a woman."[5]

There is no scientific evidence that accurate prediction can result from any of these practices listed above, although there are two opposing perspectives on that lack of verification. One is that, without clinical evidence, they must be the work of charlatans. The other is that, in your own opinion, some other-worldly concepts cannot be scientifically validated, and that their spiritual complexity places them beyond the reach of modern measurement.

If you believe that random cards can decide your future, then you would definitely reside in the latter group. Cartomancy refers to fortune-telling by shuffling a deck of cards. The first use of cards for playing games and predicting the future occurred around the mid-15th century, although they gained increased popularity in the 18th century.[6] A standard pack of 52 cards could be extended to 78 cards with a Tarot pack, which featured such luminaries as the Fool, the Juggler, the Empress, and the Lovers. When used for telling the future, the cartomancer would help the client to clarify a question. Cards would then be drawn at random, and a narrative would be woven that aligned with the cards on the table.[7,8]

In the 21st century, card readers commonly use the tarot pack to support clients in understanding aspects of their past and present life. This has some merit in itself without using cards. However, unless the card reader truly does possess predictive powers, you would derive more value by investing in some lottery tickets. The random drawing of the cards is pure chance and cannot be affected by your energy or theirs.

Palmistry has a chequered history. Partly referenced in the Book of Job, its origins are dated somewhere between the 7th and 4th centuries BC. Reading of the palm to decipher your future has been practiced in most cultures around the world.[9] Persia, Arabia, India, Nepal, China, and Tibet have all featured palmistry readings. Some of these cultures demonstrated quite different interpretations of the practice. Palm readers attest that some lines on your hand – such as the Head Line and the Heart Line – and the mounts e.g. Mars Positive and Luna mount, can be interpreted to predict your future.

A baby's hands form at very early stages, and palm readers contend that they become a record of the child's development through life. No credible scientific evidence supports this contention. Your palms may give some indication of certain dermatological conditions, but that is because of the condition of your aging skin.

Reading Crystal Balls

Perhaps you are intrigued by the thought that a clairvoyant can foretell your future. If you have ever arranged a fortune reading, a crystal ball may have contributed to the mystical environment. When a movie features a fortune teller in some form, she will often be looking into a crystal ball. Most people remember the crystal ball from movies that featured the Romani – usually referred to as gypsies – who would erect their fortune-telling tents at the side of the road when they briefly stopped on their journey. In earlier times, the practice was called scrying, which is the process of being able to see images in the crystal. The word is derived from *descry*, which means to perceive.[10]

One important feature, according to believers, was that a crystal ball was not always used just to predict the future. It could also investigate the past or find something in the present. In *The Wizard of Oz*, the Wicked Witch of The West used it as a remote viewing device to see anything she wanted to view in present time. Throughout history, clairvoyants, witches, and magicians did not only look into a crystal ball. They were also known to use mirrors, pools of water, and even the reflections in their fingernails.

The practice of fortune-telling with crystal balls began with the Celtic Druids in Gaul, Britain, and Ireland during the Iron Age (1200 BC to 500 BC), although the Romans reputedly used them soon after. In pre-1800 times, the use of crystal balls was employed by the Incas, the Chinese, the Persians, and the Egyptians amongst others. Fortune-telling was frowned upon by the Catholic Church, which considered scrying behaviour as an attraction for evil spirits.[11]

The crystal balls themselves were carved from natural crystalline stones such as amethyst, quartz, obsidian, and beryl. Some believers attested that beryl aligned with the moon's psychic energies, given that the substance had an enhanced electrical charge. Folklore indicates that the fabled magician Merlin would carry a beryl ball with him, and it would be unveiled when King Arthur needed a quick reading.

How can you differentiate between a glass ball and a crystal ball? Just touch it. Crystal balls conduct heat and can feel cooler to your fingers. The ancient Romans would sometimes hold on to a crystal ball on hot days to feel the cool surface. Many houses around the world today use a crystal ball or other crystalline objects as an ornament. Owners hang them in windows to reflect sunlight into the house. There is even an emoji of the crystal ball. Seeing your future inside any of these objects is unlikely.

The Practice of Divination

Divination is the art of using the occult, rituals, or magic to predict the future, and this practice has been a common feature in cultures world-wide. By interpreting signs, symbols, omens, and whatever message might be gleaned from the gods or angels, magicians throughout history have claimed to develop insight into future events. In ancient Greece, a priestess known as the Oracle of Delphi provided predictions by communicating with the god Apollo.[12]

Much of today's science evolved from early divinatory practices. Physics and mathematics were based upon Indian, Arabic and Pythagorean cosmological speculations. Astronomy found its origins in ancient Near Eastern and Hellenistic astrological research. Leading Renaissance scientists built their understandings on the divinatory schemes of Qabalah and hermitism as they explored the moral harmonies and direction of the universe.[13]

In many early cultures, there was often an intriguing blend between magic, religion, and science. The ancient Mesopotamians would recite magical incantations in unison with medicinal herbs to treat an illness. Many of these treatments were performed by an asipu, who was a magical arts expert. This person was often a combination of a priest, a magician, a doctor, a scribe, and a scholar.[14] The strategies for engaging in this divinatory practice had few boundaries. Incantations, hunches, sneezing, fasting, consulting the dead, and mediumistic possession all featured in the list. By the 18th century, much of this advice for predicting your life up ahead was cast in doubt.

Divination was sometimes forbidden because of political or religious beliefs. Witchcraft has featured extensively in historical journals. Criminal prosecutions were commonplace from the late 1400s until the start of the 18th century. Most of the accused were women. The Christian church often forbade any magical practice that was heretical. In the bible, Deuteronomy 18:10-11 states that: *There shall not be found among you anyone who makes his son or his daughter pass through the fire, or one who practices witchcraft, or a soothsayer, or one who interprets omens, or a sorcerer, or one who conjures spells, or a medium, or a spiritist, or one who calls up the dead.*

One common feature in divinatory rituals was the vagueness of the forecasting. You will find that this point re-occurs consistently in the world of prediction. One perspective is that it encourages the clients to develop meaning themselves from the divinatory session. Another perspective is that the magicians or seers did not really know what would happen, and so they built ambiguity into their predictions. No matter what happened in the future, they could still claim that they were correct.

An everyday belief in prediction is often fuelled by fantasy literature. In *The Lord of the Rings* (by J.R.R. Tolkien), most of the characters had some awareness of the future, whether in direct prophecies, or as a vague awareness. Galadriel sometimes used a form of water mirror to see future events. *The Golden Man*, a story from Philip K. Dick, featured an amazing human known as a precog who could predict the future up to the time of his death. In the movie *The Minority Report*, the precogs could sense the future. This sometimes led to the arrest of people even before they had committed a crime.[14]

What about the ability to read someone's thinking? Other than the fictitious characters in today's fantasy reading genre, there was rarely, if ever, a human being who could genuinely read someone's mind. In most cases, it was a combination of psychological manipulation, misdirection, and verbal cues. Given the enthusiastic response of most modern audiences to a successful mentalist, we obviously love to be entertained with an alleged mind-reading act. Just imagine the consternation you would have created in the 1700s with your mentalist performance at the local village fair. These acts were mere entertainment, and rarely if ever a demonstration of true magical or predictive ability.

Astrology and Dreams

There was magic in the heavens as well. The sky in ancient times was today's version of a large TV screen full of moving images. Astrology (not to be confused with astronomy, the science of everything in the Universe outside Earth's atmosphere) reveals itself in the myriad daily horoscopes available around the world. In Gallop polls taken over 25 years in Britain, Canada, and the US, around 25% of people stated that they do believe in horoscopes, and that they take these cleverly worded predictions seriously.[15] Another large cohort (and I am one of this group) scan them for the mere fun of it, and immediately forget what they have contained.

If you have it, please send me verifiable scientific proof that astrology has evidence-based merit. I could not find any, and with an open mind, I looked everywhere. Admittedly, belief in astrology has persevered through the centuries, which demonstrates that it has some form of mass appeal. In a 1943 letter, Einstein reputedly wrote that *"I fully agree with you concerning the pseudo-science of astrology. The interesting point is that this kind of superstition is so tenacious that it could persist through so many centuries."*[16]

As usual, there is some small degree of truth hidden in the details. Astrology is based upon the belief that magnetic fields generated by heavenly bodies will influence your life. There is some evidence that the magnetic fields from the earth itself affect us. Other evidence indicates that the moon and sun also have some influence on us. To claim that you were born under the star sign Leo, and that it affects aspects of your life at age 42, is a very long draw of the truth bow. The stars in that constellation are on average about 1535 light years away (approximately 14,583 trillion kms; or 14,530,000,000,000,000,000 km).

There may be one partial exception to the many pseudosciences that incorrectly claim to predict the future. Dream interpretation has been rigorously analysed, and clinical evidence exists that your dreams can help you to understand what is happening in your life.[17] No scientific data is available that conclusively proves your dreams can foretell the future. One likely explanation about this form of prediction is that dreams are related to past experiences, and that these can sometimes combine to create a dream that gives an indication it will occur up ahead.

Once people have dreamed the event, they may engage – subconsciously or otherwise – in creating the reality. As well, one out of every one hundred dreams may eventually come true, although this is more likely because of pure chance than prophecy. On rare occasions, the dreamer really may prophesy the future. Several friends have sought to convince me that they occasionally experience daytime visions of events up ahead. This is obviously only hearsay, and not scientifically valid proof. Many earlier visionaries claimed that their dreams could foretell the future.

Historical Visionaries

One well-known visionary was the 16th century French astrologer Michel de Nostredame. More commonly known as Nostradamus in the English world, he published a book in 1555 called *Les Prophéties*. It has never been out of print, which admittedly most authors today cannot claim. This book contained 942 poetic prophetic quatrains, which reputedly predicted many events right up to the 21st century. It has been reported that he dictated them to his secretary while he overdosed on nutmeg, which can cause hallucinations.[18] This is one compelling explanation for the eccentricity of the prophecies.

An industry of sorts has arisen out of decoding his prophecies. You will probably believe what you want to believe. I admit that I dismiss most of them, but not all. However, given that he wrote 942 of them, surely there is a random chance that some of them must eventuate. As well, we all can be excellent in hindsight, which allows us to reinterpret an event according to the words in the prophecy... after the event.

One specific quatrain may prove to be interesting. In quatrain 10/22, he said: *"Because they disapproved of his divorce / A man who later they considered unworthy / The People will force out the King of the islands / A Man will replace who never expected to be king."* According to some scholars, this means that Prince Harry may eventually become King.[19] If that eventuates, I will re-evaluate everything I believe about Nostradamus.

Vangeliya Pandeva Gushterova (1911 – 1996), commonly known as Baba Vanga, was blind since early childhood, and spent most of her life in the Kozhuh mountains in Bulgaria. She reputedly foretold the breakup of the Soviet Union, the Chernobyl disaster, and the September 11th attacks on the Twin Towers. All of these foretellings were circumstantial, and no documentary written evidence corroborates these predictions being made before the events.[20]

You can trust the Greek legends to provide an intriguing example of a prophet. Cassandra was the daughter of King Priam of Troy and his wife Hecabe, and was considered to be one of the most beautiful women in the city. The god Apollo attempted to woo her, and while doing so, gave her the gift of prophecy. However, she spurned his advances. In seeking retribution, Apollo deemed that she would retain the gift, but that no-one would ever believe what she said. The blessing became a curse.

Although it was probably an imaginary fable, Troy was eventually over-run after a 10-year siege by Greek soldiers who hid in a giant wooden horse outside the gates. Although Cassandra warned the city of the coming calamity, the citizens again refused to believe her. We today refer to the Cassandra Complex, in which some people might accurately predict the future, and yet no one will believe them.[21]

Given its historical path, what can we tell about the art of prediction up until now? After all, there has been no shortage of farcical and creative strategies over the past few thousand years. No scientific evidence conclusively proves that these options have ever been able to accurately foresee what is up ahead. The quatrains written by Nostradamus have some inconclusive credence, although the majority cannot be substantiated. And yet, some predictive practices such as weather forecasting and astronomical studies were reasonably accurate for their time.

Why therefore would we seek to understand how past civilizations engaged in prediction? There are two worthwhile reasons. One is that earlier predictive practices such as the mapping of astronomical bodies were created without help from modern technologies. This demonstrates an ingenuity that can even instruct our practices today. The second is that many age-old predictive practices formed the basis for our modern beliefs. Researching the history of those practices – whether mathematical, philosophical, mystical or scientific – can enhance our understanding of present-day society.

From a modern perspective, what does prediction really mean? How accurately can we determine the likelihood of something occurring up ahead? In Chapter 2, we'll explore some different perspectives on prediction, and clarify the astonishing value of forecasting the future.

Predictive Exercises to Expand Your Predictive Ability

- Write your own star sign prophecies. Don't just guess. Base them upon the strong probability they will occur. Take note whether you're more likely to then place them into action.

- Set up a role-playing game where you assume the perspective of a person from a past civilization. Make decisions based on their knowledge and beliefs.

- List any superstitions you have. Ask yourself: Do they enhance my life? Can I really predict with them? Or am I deluding myself? If you can, clarify a genuine example of accurate prediction that was derived from a superstition.

- Invent a superstition and practise it for a week. Document any changes in your behaviour or outcomes. Reflect on how beliefs influence your actions and perceptions of the future.

- Write a short story set 20 years in the future based on current trends in technology, society, and politics. Share it with others and gather feedback on the plausibility of your predictions.

- Research historical predictions that did not come true. Analyze why they failed and identify any common pitfalls or biases that led to those incorrect forecasts.

- Write a journal of your night (or day) dreams for several weeks, and determine whether any do indeed predict aspects of your future.

- Develop your own predictive versions of a short story that best fits with each Tarot card; then weave those stories together from a random fall of the cards.

- Choose a positive historical event and study the factors leading up to it. Identify patterns and triggers that could be applied to predict similar future events.

CHAPTER 1 IN SUMMARY

- Early civilizations relied strongly on superstitions and rituals to influence or predict the future, reflecting a deep-seated fear of the unknown and a desire for control.

- Throughout history, various methods such as crystal ball reading, palmistry, and astrology were used to foresee the future, though few have been scientifically validated.

- A common feature in historical predictions was their vagueness, allowing for flexible interpretation and thus perpetuating belief in their accuracy.

- Many of today's beliefs and practices have roots in ancient predictive methods, highlighting the lasting impact of historical approaches to forecasting the future.

- Figures like Nostradamus and fictional characters like Cassandra exemplify humanity's enduring fascination with prophecy, despite the often dubious nature of their predictions.

CHAPTER 2: UNDERSTANDING PREDICTION

Predicition is the essence of intelligence.

YANN LeCUN

Your Brain is a Predictive Entity

When is the last time you caught a ball that was thrown towards you? What your brain does as you prepare to catch the ball is highly predictive.[1] You not only watch the ball in flight, but you also predict where it is going. You use a series of visual cues, and adjust your hand position to catch the ball. While the ball is moving through the air, you monitor its flight path, and minutely update your predictions of its arrival time. And when you catch the ball, your brain confirms the accuracy of your prediction. If you catch a ball often, your brain continually adapts and improves the way that you engage in the activity.

Your brain analyses what might happen in your near future with many everyday experiences.[2] Strong predictive ability indicates higher intelligence, given that you are more likely to solve problems. You analyse relevant data by using some predictive factors, and then determine the probability of what might occur up ahead. Those predictive factors can include the recent history of similar events, the age of people involved, cultural backgrounds, or who is directly involved.

Driving a car requires predictive perception of what other drivers are about to do. Without this ability, you are less likely to survive the drive. When you play most sports, you often predict what the other player or team is likely to do in response to your own actions. Some people – including me – ponder the next few minutes, hours and days of life beforehand, and then pre-plan how to maximise on every opportunity.

Matthew Syed maintains that "Prediction is central to pretty much every decision we make, whether at work, or in everyday life."[3] In 'Superforecasting: The Art and Science of Prediction', Philip Tetlock and Dan Gardner wrote that we are all forecasters through necessity: "When we think about changing jobs, getting married, buying a home, making an investment, launching a product, or retiring, we decide based on how we expect the future will unfold."[4] Your ability to analyse your future can create a tangible advantage for you. To help you with developing that advantage, let's clarify what prediction really is.

What IS Prediction?

Prediction (Latin *prae*, "before", and *dicere*, "to say") is a statement about something before it happens.[5] There are various synonyms for prediction. Forecast. Foretell. Prophesise. Prescience. Plan. Prognosticate. Create. Precognition. Foreshadow. My interpretation is that they are all variations on determining what might happen up ahead. Forecasting and prediction are the two most common overlapping terms, and I will use them interchangeably throughout this book. Some futurists consider forecasting to be based upon data analysis, while prediction might rely more on personal judgement.

Various acronyms are used to predict the future. Not many are heartening. VUCA is well-known, and stands for Volatility, Uncertainty, Complexity and Ambiguity. Each of these qualities attest that the future environment will be dynamic, uncontrollable, disrupted, and in a constant state of flux. BANI is a more negative concept, and stands for Brittle, Anxious, Nonlinear and Incomprehensible.

Prediction can be complex. We are dealing with human choice, probably the most significant variable on this planet. No amount of data analysis can fully unravel what people might decide to do on a whim in the future. While a specific prediction might be made today, it may become wildly inaccurate by the time that large numbers of people have become involved. Because of this, the complex factors required to program a computer simulation of a future event can be infinite.

While predictions can be made for events that are days or months ahead, the most common form of predicting the future occurs just a few seconds ahead. This is to your lifetime advantage. You predict what will most likely occur when you engage with certain people in a specific way. You predict the content of an article after reading the first sentence. You predict the quality of a movie within the first minute. You predict how your children will react to specific challenges, which helps you to positively resolve an issue with them. Whether it is in the short or the long term, there are three perspectives on what we mean by prediction.

Three Perspectives on Prediction

The first perspective is that you cannot predict anything that occurs by chance. An event that is pure luck - such as the Lotto winning numbers for this Saturday night - cannot be determined beforehand. Black Swan events - such as earthquakes and sudden accidents - occur out of the blue, and no amount of scientific analysis can forecast those disasters.

Nassim Nicholas Taleb, author of *The Black Swan*,[6] writes about the critical importance of uncertainty and randomness. For him, there are too many variables involved in most events to ever hope to foresee with any certainty. Predicting long-term geopolitical events is fraught with dilemma, given the complex circumstances involved with those situations. You can expect the unexpected.

The second perspective gives us more cause for thought. It is when you extensively research the probability of an event occurring, while taking account of every possible factor, and then adding some human wisdom and insight. Think of weather reports, which offer you a scientifically valid percentage chance of rain e.g., 70%. Such a prediction combines some meteorological analysis with the foresight of an accomplished weather forecaster.

All predictions about any aspect of life fit somewhere between a highly unlikely and a definite probability of happening. In contrast to Black Swan events (which are near-impossible to predict), a White Swan occurrence is very predictable, and will have a high probability of taking place. COVID-19 might be one example. A grey swan event has a low probability of happening, although its impact will be significant. The 9/11 attacks on the US would be one example of this descriptor.[7]

The third perspective on prediction is when you decide to accomplish something, and you are strongly committed to making it happen. Let's call it self-determinism. For you, the future is not something to predict. It is something you shape. As Peter Drucker said: *"The best way to predict the future is to create it."* For tomorrow, you consciously plan to eat a healthy breakfast, smile at your children when they wake up, take public transport to work, complete your latest mini-project, and exercise in the local gym when you finish work. There will sometimes be circumstances that prevent you from doing some of these, but committing to them increases the probability of them occurring.

When you engage in a healthy lifestyle, you are more likely to enjoy a healthier future. Saving and investing your money will prepare you for a time when you may be less capable of earning it yourself. Raising your children is perhaps the ultimate example of creating the future. Filling their lives with an entire series of protective factors - unconditional love, good friends, sport, community activities - is more likely to lead to them becoming self-confident and loving adults. Your future self will be grateful for these efforts by your present self.

An opposing theory to self-determinism is known as pre-determinism, in which you believe your future has been mapped out through some universal design. Some might call it fate. If you believe this, then there is nothing you can do to influence your life up ahead, let alone the lives of your family or friends. I humbly recommend that you stop reading this book if that is your perspective. To you, every accident was meant to happen. Every stroke of good luck was only through divine intervention. You are convinced that you have little agency over your life, and that you are merely following a path that has been pre-ordained for you. I pay respect to your belief. I just don't agree with it, given everything that I have researched about personal capacity to influence one's future.

There is an overlap between these two beliefs. A balance between self- and pre-determinism might be referred to as opportunity, in which you stay open to possibilities that the universe presents to you. When such an opportunity arises, you would make every effort to respond to that chance. This mid-point belief accepts that while the universal flow is all-powerful and difficult to change, you can still occasionally influence its direction. Whichever belief you hold, you will soon come to understand that prediction is not infallible. Some forecasts will turn out to be accurate, while others will end up being wildly incorrect.

Good and Bad Predictions

History is littered with both impressive and poor predictions. Some geniuses from the past have provided us with examples of accurate foretelling. In 1865, French author Jules Verne wrote of three astronauts in a spacecraft called Apollo, taking off from Florida and landing on the moon after experiencing weightlessness in space. Nikola Tesla predicted the appearance of Wi-Fi and mobile phones. In a 1909 interview, he foresaw the creation of the world wide web at least 60 years before it became a reality. Tesla said: *'It will soon be possible to transmit wireless messages all over the world so simply that any individual can own and operate his own apparatus.'* [8]

In more modern times, US baseball statistician, election analyst and professional poker player Nate Silver employed statistical models to predict the outcome of the 2008 United States presidential election. The race between John McCain and Barack Obama was considered by many pundits and pollsters to be close, and Silver was criticised for his perceived strong support for Obama. He combined historical trends, demographic data, and a variety of polls to correctly predict the result in 49 out of 50 states. [9]

His 2012 results were even more impressive, when all 50 states were correctly forecast. Part of his success can be attributed to methodologies that account for uncertainties that influence outcomes e.g., a poll may only include people who own landline phones. Silver writes that prediction is certainly not failsafe, and that it connects "subjective and objective reality". His website at FiveThirtyEight [10] has become a respected source for election predictions.

Others will be forever remembered for their predictive ineptitude. IBM president Thomas Watson reputedly claimed in 1943 that "I think there is a world market for maybe *five computers.*" December 21st, 2012, was prophesised as the end of the world because of the Mayan Apocalypse. Sales of survival kits and shelters spiked significantly. We are still here. Most religions share legends about the origins of their faith, along with apocalyptic narrative of what will happen if we collectively sin. Numerous predictions of floods, fire, and famine litter the historical landscape. It makes you wonder if you would rather not know what is up ahead. Is it possible that ignorance about the future might be bliss?

Some Pitfalls of Prediction

It's all very well to predict some of your future, but would you want to know everything? This is unlikely for most. I do not want to know when I'm going to die, and I would rather not find out the same thing about any family members or friends. You may have other events you would prefer not to know. To intuit every single moment of your future existence would be a strange life. Imagine waking up each day and knowing everything that was going to happen, without exception. Shades of Groundhog Day.

In any case, it would be impossible for every person to know everything about the future. Everyone would win Lotto. Everyone (theoretically) would live an amazing and successful life, as long as they made the effort to place their forecast information into beneficial practice. There are many other philosophical implications with prediction.

Just making a prediction can alter a possible future event. Your doctor might warn you that your smoking is narrowing your veins and arteries, which is forcing your heart to work much harder. Your life expectancy is likely to be reduced. Because of this dire warning, you give up smoking, which then generally lowers the risk predicted by your doctor. A forecast outbreak of the flu by government authorities might compel a large percentage of the population to arrange a vaccine. This alters the predicted numbers who would have otherwise contracted the flu.

An ability to predict the future can become a philosophical tangle. If you knew your future at a 100% certainty, wouldn't you attempt to alter some of it before you arrived at that point? If you predicted an accident, then surely you would avoid it or at least, minimise on its effects. Your intervention might then change the prediction or delete it altogether. Does that mean it was never going to happen in the first place? If so, the prediction then was incorrect. And if you foresee a positive result, might you then become complacent and not bother to work at making it happen? In many cases, optimistic things happen for you because you make the effort to create the result. The prediction might again become incorrect.

Another criticism of having predictive ability is that you might think only about your future self, and neglect to experience life right now. That is not necessarily the case. I find that I can often conjecture about the future, while still remaining centred in the present. I stay future-aware in everyday life without being obsessed about it. Friends admit to me that worry becomes their constant companion when their thinking is predominantly about the future. My response is that you are better off understanding what might occur up ahead, as it gives you the opportunity to prepare more effectively.

Why Prediction Has Merit

While there may be some things you do not wish to know in the future, the core premise of this book is that accurate prediction can create many benefits for you. Those benefits can involve events over the next few seconds, or long-term experiences over succeeding years that will affect your physical, mental, financial and spiritual health. Here are nine categories in which predictive analysis may have merit for you:

Health: We are now moving from your 'medical history' to your 'medical future'. Health specialists have always known that prevention is preferable to the cure. DNA tests (see Chapter 4) can provide some indications about future issues for your health. Public awareness about the barometer of poor future health is now stronger than ever, given that Dr Google has arrived on the scene. Many actions and choices adopted at the right times can create a life less prone to disease.

Stability: When you perceive that the world is volatile and outside your control, you can too easily feel unsettled and even fearful about what is up ahead (see Chapter 10). Predictive capability can generate an opposing mindset. At least you will be determining aspects of the future yourself, rather than letting others dictate their thinking to you. Your locus of control moves from being outside you, to creating stability within you. Conversely, an external locus of control means that you will be held captive by horrendous media sensationalism, or by the relative who can only talk about the miseries of the future.

Relationships: While there is little evidence-based reality to the 7-year itch, the ability to predict the future of a possible relationship will save you endless heartache. Gut feeling matters in the very early stages. In the longer term, looking for indicators of concern can save your present relationship. Instead of slowly sliding downwards, you can implement some interventions that redress the predicted issues.

Intuition: Trusting your intuition supports your predictive efforts (see Chapter 7). Even an overwhelming amount of data cannot create the prediction all by itself. The data needs to be combined with gut feeling, something that robots are unlikely to do for a while yet. Throughout human history, we have struggled to accept the worth of intuition. It has been denigrated because there is little scientific validity, and yet your premonition about what is about to happen may be a valid predictive tool.

Financial security is an obvious benefit of predictive ability. We are not talking about winning the lottery here, but instead, evaluating and acting on some strong research that can steadily lead to a comfortable standard of living. The only way you might (repeat: might) benefit from gambling is by approaching it professionally (see Chapter 6). Earning money, and then investing it in your chosen financial category – whether it be shares, real estate, or rare art – is the most obvious approach. This investing can be based upon professional advice, as well as your gut instinct.

Thinking. Your thinking capacity benefits strongly from engaging in prediction. Through the rest of this book, you will discover worthwhile strategies for analysing what is up ahead. Those strategies all need you to think deeply. You are likely to make better informed decisions, especially when you are researching the probabilities and trends of specific events. At the very least, it will be great exercise for your brain (see Chapter 9).

Sport. You will play more effective sport. There are at least two supportive points to this. The first is the long-term, in which you accurately predict and then action your results (or your team's) over a full season or even longer. The second is that most sport requires you to predict what your opponent is about to do. The better you are at a sport, the more effective you will be at rapidly calculating what your opponent will do next. A split-second advantage makes all the difference. See Chapter 4 for more detail.

Gratitude. When you engage in open-minded research, you may just discover that the world is not all gloom and despair. Billions of people right now are modelling what has been successfully accomplished all through history. They are resilient, adaptable and capable of creating solutions to difficult problems. You may even express gratitude for what has already been achieved in our history. You can see the possibilities (and yes, the perils) up ahead, as outlined in Chapter 10. Forewarned is forearmed, and having confidence in your predictive ability can lead to effective interventions for you, and even for the rest of the human race.

Predictive analysis does not just have benefit for you and other individuals. It also has merit for many of our important institutions around the world. Here are some predictive What-ifs for some of those institutions:

Predictive Medicine. What if remote personalised predictive warnings – especially with wearable tech - could be provided to anyone who was compromising the health of their own body (or brain)? Prevention would be better than the cure (and it often costs less).

Predictive Policing. What if the probabilities for most accidents and crimes could be predicted before the event occurred? More policing would then focus on prevention rather than intervention. Roll on Minority Report.

Predictive Politics. What if all politicians and policy writers could be privy to computer simulations that would accurately predict geo-political and climate change consequences that would result from their decisions?

Predictive Business. What if businesses could develop predictive algorithms that clarify clear patterns in their operation, and then maximise on profits that are gained by applying the patterns with updated data?

Predictive Education. What if educators could accurately predict the capabilities that will be needed by young people in the latter half of this century, and then focus today's learning on the development of those capabilities?

Predictive Environments. What if the results from predictive analysis of the world's faltering environment could guide how to apply the most viable interventions?

Predictive Religion. What if all religions engaged in predictive analysis to determine the ethical beliefs needed in a future world (eg with online civility), and then provided the social structures that would develop those beliefs?

Predictive Non-profits. What if non-profit charities could accurately predict the needs required by struggling citizens in the future, and then plan ahead to provide that support?

Given that predictive analysis has many benefits, where is the best place for you to start? Think in terms of probability. Let's explore that in Chapter 3.

Predictive Exercises to Expand Your Predictive Ability

- Every morning, write down three predictions about your day. At the end of the day, review and analyze the accuracy of your predictions, considering why they were right or wrong.

- Take note of how often you are predicting the future each day, whether it is at work, in sport, or during your interactions with your family or friends. Become clearly aware that you are doing that forecasting, rather than only doing it subconsciously.

- Engage in discussions with friends or colleagues about future events. Make predictions about upcoming sports games, political events, or technological advancements. Compare notes and discuss the reasoning behind each prediction.

- Identify a well-known expert in a field you are interested in (e.g., Nate Silver in election forecasting). Study their methods and try to emulate their approach in your predictions.

- Make some short-term and long-term predictions for an industry or profession in which you work. Write them down, and refer back to them in a year from now.

- If you drive a car, take note of how far ahead in time you can predict the actions of nearby drivers. Check if they give small signals of what they are about to do eg moving slightly into another lane, even before activating their indicators.

- Most people think of prediction as something that will happen hours or days ahead. As a variation, learn to become more aware of what might occur just 5 seconds ahead.

CHAPTER 2 IN SUMMARY

- Our brains constantly engage in prediction, whether in catching a ball or driving a car, using past experiences and real-time data to anticipate future events, which is fundamental to intelligence and decision-making.

- Predictions can range from the unpredictable (e.g., Black Swan events) to highly probable outcomes, and even self-determined futures where individuals shape their own outcomes.

- History offers examples of both successful and failed predictions, illustrating the complexities and uncertainties involved in forecasting future events.

- The ability to predict can alter future events and raises intriguing questions about the influence we have on your own future lives.

- Prediction has significant value across various fields such as health, finance, education, and politics, enabling better planning, prevention, and adaptation to future challenges.

CHAPTER 3: THE SECRETS OF PROBABILITY

We may not be able to get certainty, but we can get probability, and half a loaf is better than no bread.

C.S. LEWIS

Possibility versus Probability

The *possibility* exists that you might win the Lotto. The *probability* of you winning is 8,145,060 to one for regular lotteries (e.g. choosing 6 numbers from 45),[1] and approximately 292.2 million to one for the US Powerball. Possibility is a generic sense of something happening in the future. Probability offers a percentage chance that it will occur, and this can require extensive prior research.

A percentage probability is often better than an emphatic Yes or No on whether something will take place up ahead. It indicates caution, which is needed for nearly all predictions. This also provides an excuse if your prediction does not eventuate. After all, predicting a 70% chance of your sports team winning also means a 30% chance that it won't. Either way, you're covered (unless you bet on it).

When you are determining the probability of a specific result, make sure that you are not influenced by 'wishful thinking'. Just because you hope that your sports team will win has no effect on its eventual results. It is all too easy for your thinking to be affected by your bias. Hope has significant merit in your personal life, and it is critical that you approach your future with some degree of optimism. But when you use hope to accurately predict the future outside of yourself, it has less relevance. Using hope indicates a confirmation bias, in which you only value information that aligns with your hoped-for prediction.

Everything that can be predicted rests on a spectrum between 0% and 100%. This is a critical point about prediction, and forms the basis for most forecasting efforts. Very few everyday events are at either end of the continuum. The sun to rise tomorrow morning? 100%. Outer space aliens knocking on your front door right now? You may disagree, but I'll rate that one at 0%. And everything else fits inbetween those two extremes. The issue then becomes how we can calculate the probability percentage for a specific forecast.

The technology for determining a probability has improved massively. Fifty years ago, weather forecasting was reasonably accurate only one day ahead. The same standards now apply to a five-day forecast. Probability indicators are all around you. Modern production lines can include technology that calculates machine downtime, and signals it is time for maintenance. Some cars do likewise. What you pay for insurance is based upon algorithms that determine the chances of something happening to you or the insured item. A huge range of data points contribute to the final probability determined by a computer.

Hindcasting can contribute to the calculation of a probability. Forecasting is done before an event occurs. Hindcasting is applied after the event. Large databases are established that contain every possible factor evident in the previous event e.g. storms, and the data will include the accuracy of predictions that were made before the event. After another storm occurs, any new observations are added to that database. The experts improve their future forecasts by studying what previously was incorrectly predicted, and learning from those mistakes.[2]

Predictive Analytics

How else is probability calculated? In a world that is overwhelmed with Big Data, the process of predictive analytics helps organisations to sift through all available information, and to determine accurate probabilities. That forecasting can include whether their stock reserves will need to be adjusted, or whether customer tastes might soon change, or how cashflow might be compromised by a national holiday. With astute analysis, overall cashflow can be calculated. This changes how any business thinks of income. Rather than hoping to make money, they accurately calculate the amount they will make.

Predictive analysis generally requires a step-by-step process for an organisation. Here is one possible process:

1. Clarify what specifically is being predicted eg the sales of rooftop solar panels in your city over the next year

2. Collate data from several sources such as databases, sensors, and web sources. This may include localised surveys, feedback from social media, and future trending based upon data from the past year

3. Choose the contextual data that is most relevant to the eventual prediction eg the number of people who have indicated they are considering a purchase of solar panels

4. Construct an accurate predictive model by using statistics and machine learning, which is calculated with analytic software. Amazon Forecast, Amazon QuickSight, Adobe Analytics, or SAP Analytics Cloud are just some that are available

5. Integrate the predicted model with the real-life environment

Given the complexity of systems today – think of managing loads in your region's electricity grid – this predictive analysis is undertaken by large-scale technology. An online search for 'predictive analytics software' will uncover a range of companies willing to help you out. If that seems irrelevant to you, rest assured that organisations such as weather bureaus, car manufacturers and power companies are constantly doing this to provide a better quality of life for you.

If you have ever arranged to pick up an arriving relative at the airport, you likely have used the airline's AI predictive software to determine the arrival time. This can take into account the earlier departure time, weather during the flight, and even delays at the arrival airport's disembarking gates. Your Amazon order sometimes even arrives on the day of ordering, thanks to AI prediction. As well as assessing the likely daily demand for a specific product, it also takes traffic conditions into account, and can calculate the most efficient delivery method.

The strength – and the weakness – of predictive analysis is that it is sometimes based upon historical data. If there is little variance between past and future events, then it may work. However, this approach may fail to account for uncertain influences – such as rapid updates in available technology - that might appear with little forewarning in the future. Perhaps the most problematic influence is human behaviour. Forecasting a future response from one person, let alone one million, is complex.

To help with this, an AI model from MIT predicts human behaviour with strong accuracy. Like most systems up to now, it predominantly develops its forecasts by analysing past behaviours. Unlike most earlier systems, it has been coded to understand and respond to human decision-making and irrationality.[3] This indicates that, within a few years, AI that can comprehend human fallibility will engage in predictive tasks far better than humans. However, one critical element will still be required. Whatever predictive system is employed, we will need to find data that is as precise and as updated as possible.

Finding The Best Data

Making an important decision can be problematic. Which job should I accept? What small business would best suit my personality if I wanted to start one? Who is the better partner for me? One course of action is to use predictive analysis, and ascertain the most likely consequences for any decision. Two issues will determine the effectiveness of those analyses. The first is the incisiveness of your predictive question/s. You might ask: What really matters the most? What's the one big idea in all of this? What do I most want to find out? What one issue would make things better for me up ahead?

Although open-ended questions (i.e. ones with many possible answers) have merit, you need to be explicit with what you ask for here. If matching up with an inspiring partner is what you seek, you may ask: What social characteristics do I need to steadily develop in order to attract the type of person who appeals to me? If you have a side hustle with an online clothing store, the question may be: What specific items of clothing are most likely to be in fashion in eight months from now?

The second issue affecting the quality of your prediction is the worth of your data, both quantitative (evidence-based) and qualitative (subjective).

Acquiring information today is a vastly different experience to a hundred years ago. Researchers once relied on a visit to a public library, or they leafed through a set of the *Encyclopaedia Britannica*. This expensive collection of books was first released in Edinburgh around 1768.

Over the next two centuries, it became the most common research tool on the planet. However, the online world rapidly surpassed this paper-based resource, and the 2010 version of 32 volumes and 32 640 pages was the last printed edition.[4] This was a turning point in the aggregation of data. Printed material rapidly became outdated very quickly, whereas an online search for data about your possible prediction can find information that has been updated that same morning.

The downside to conducting research online is that it is a bottomless pit. At least 3.7 million videos are uploaded to YouTube every day. That equates to over 270,000 hours of new viewing pleasure every 24 hours.[5] The upside to your research is that it is vastly easier to find what you need with decent search strategies. Type 'How to Search Online' into your preferred search engine, and the further ideas you find - such as how to write your search query - can save you many hours in research time.

To widen your search results, vary the search engine you use. While up to 92% of all searches might occur on google[6], there are many other feasible options. Duck Duck Go has merit as an alternative search engine, given that its anti-tracking policy leads to results that are not biased towards your previous searches. The concept of a needle in a haystack does not even begin to demonstrate the complexity of finding worthwhile online data. Around 90% of the sum total of global data in history has been developed just in the past two years.

How can you deal with this overwhelming amount of data? Clay Shirky maintains that it's not info overload that is the issue. It's filter failure.[7] Learn how to curate, or to filter, the information that you need. Set up your own newsfeed or news app. Rather than doing all the searching yourself, tap into the expertise of others and find out what they have discovered. Most social media apps work on this premise. The Flipboard app allows you to search the newsfeeds of other people who are knowledgeable in their field. Flipboard offers categories such as General News, Sport, Music, Food, Technology, and The Future.

Large language models (LLMs) such as ChatGPT have revolutionised the writing of a data report, or blogs, or recipes, or music. While previously you needed to search for the information, and then organise it into a coherent series of paragraphs, the LLM now does it for you. But beware the limitations. For starters, this AI occasionally produces some variations on the truth. Cross-referencing is critical when you use ChatGPT. Be careful of relying too much on this technology to do the work for you. Having to write something yourself is still important for maintaining your brain's functionality.

Whether you use a search engine, or an LLM, the quality of the search query will generally lead to more effective results. Queries that are succinct, contain fewer prepositions (e.g. beneath, upon, with, around), and have relevant keywords will more often find what you want. If you are feeling cognitively lazy, then ask ChatGPT for the most effective question that you could use for your topic. You might ask: What's the most effective question to ask when I'm looking for trends in technology? Its answer might be: What are the emerging trends that are shaping the future of technology? Then use that question to prompt the best possible answer.

Prompt engineering[8] refers to the skillset that improves your search inputs. As an example, you might begin your prompt by telling Chat-GTP to act as an experienced data analyst for online clothing sales. You would then include the specific data that you want it to analyse. Another option? Instruct the AI to outline the steps it uses to find its answer, and this may give you further worthwhile search options. Or perhaps you might ask ChatGTP to criticise its own report. After it has written a report for you, prompt it to critique and then rewrite it by using its own suggestions. Your time spent on becoming a competent prompt engineer can lead to improved research and predictive analysis.

Bayes' Theorem

One theory over 270 years old can help you to calculate the probability of future events. In 1763, English Presbyterian minister and philosopher Thomas Bayes published 'An Essay towards solving a Problem in the Doctrine of Chances'. One of his legacies is that he gave his book a complicated title. Another much more important legacy is known as Bayes' theorem. Sometimes called Bayes' law or rule, it calculates the probability of an event, based upon prior knowledge of any conditions that might be related to the event.[9]

This law states that your new prediction depends upon your prior belief, but multiplied by the diagnostic value of any new data. Bayes' Theorem is a mathematical formula for determining conditional probability. This is the likelihood of any outcome occurring, which in turn is based on previous outcomes that have occurred in similar situations. Bayes' rule allows you to also update existing predictions, given new evidence.

In the modern world, it can be applied in any circumstance where prediction provides a distinct advantage. This can include financial issues, such as the calculating of risk evaluation, or for medical analysis requiring an accurate probability that a patient has a specific illness. Until recently, Bayes' theorem was not often used to complete more complex transactions because of the calculation capacity that was needed. Complicated examples such as cancer rates or defective machinery rates now can be calculated with an online Bayes' Theorem calculator.[10]

There are, however, many everyday situations that can be resolved with the theorem. As one simple example, you might draw a card from a deck of 52 cards. The probability of that card being a King is 4 in 52, or 1 in 13 (about 7.7%). However, if you were provided with new information (eg that your card is a face card), the calculated chances that your card is a King increase markedly. Because there are 12 face cards in total, the probability is now 4 in 12, or 1 in 3 (33.3%).[11]

Laplace's Demon

Another well-known mathematician called Pierre-Simon Laplace followed on from Bayes' work fifty years later. Predictive efforts in the early 19th century were often based upon pre-determinism, in which it was believed that the Universe had already decided what will happen to you up ahead. People conjectured that an infinite number of dominoes were pre-set to fall at an appointed time. Supporters of this theory claimed that to predict the next month, you merely had to analyse the past few months. This can sometimes work, but in a rapidly changing 21st century, it is less likely to be successful.

If you do believe that everything in your life has been pre-determined, then you will appreciate a thought experiment proposed by Laplace in 1814. He asked his readers to imagine the most intelligent and all-knowing entity in the Universe.[12] This entity would be aware of everything that had ever happened anywhere in history. Driven by the common belief in pre-determinism, Laplace proposed that this entity would also be able to predict everything in the future. This all-knowing system later became known as Laplace's Demon. Why a Demon? Partly because of the power it would unleash, and the unease it would create.

In the next few decades, we will experience more advanced versions of Laplace's Demon. Computer simulations can already approximate specific future events, including possible changes to weather patterns from climate change. The dilemma is that it is still difficult to account for every possible factor. Advances in quantum computing will enable scientists to include a near-infinite set of factors. Creating a gaming environment that mirrors real-life issues will be commonplace, and will feel more relevant. Think of melting glaciers, or earthquake damage, or experiencing the implications of major geo-political decisions.

Imagine the demand for advanced computer games - populated by real-life characters, including you - that could accurately create a future scenario. The study of Games Theory, and the resultant computer game simulations, helps participants to understand what may occur in the future. In most cases, games are played between two players or groups who are not permitted to communicate with each other during the game. While predominantly used in business and economics, game theory can also be applied in carbon trading, sustainable development, and environmental science.[13]

At the University of Texas in Arizona, Prof Karen Willcox has explored the possibility of digital twins.[14] These can be created for virtually any object or concept, including engines, aircraft, your own life, or even the planet's environment. While the concept, entity or object in question exists in real-time, the digital twin exists in a virtual space. These virtual simulations can accurately analyse what will occur in the future of the real object, whether it be with aircraft engines, or even your own health. This process is known as data assimilation, in which the data gleaned from the virtual twin can accurately predict what is likely to occur with the real device.

Another version of a digital twin involves you interacting with your online avatar.[15] This twin of yours might eventually do some / most of your work for you, and could attend online meetings, write reports, and even give a presentation of your material. Some users marvel at the prospect, and delight in the thought that their twin will do their work for them. However, the ethical pitfalls are complex. Will you be legally responsible for a mistake made by your digital twin? What of the privacy and security issues involved? Please be aware that to create one of these twins, you need to populate it with lots of your own personal and professional history.

Added to the mix will be the possibility of sentient technology, in which a device truly is capable of near-human thought and feeling. This might supplant our own ingenuity and intuition. Such a device may be the closest we will get to technologically predicting the future. Laplace would have been fascinated by these possibilities. Once this high tech is integrated with other effective processes already outlined in this book, the calculation of probabilities will improve further.

Spotting the Trends

Social media influencers set out to create trends. They introduce a novel new process, and hope that their followers then share that process far and wide. One effective TikTok trend is simply called Rating. The Tik Tok user videos her family or friends engaging in everyday activities, and then gives a rating on that activity. Not cleaning up after lunch might incur a 1, while giving her a chocolate might be a 10. This process can be used with politicians, your pet dog, or even yourself.[16] Why would you ever take part in any of these trends? Partly because you like to be a member of the group who engage in the same experience.

Instead of creating a trend, other trend analysers carefully observe people in everyday life, and predict the probabilities for what will happen in the near future. Endless sites will provide these observations for you, and the topics can include fashion, transport, travel, beauty, business, and even weird gifts, such as sending your ex-partner a cactus. A trend is basically what is popular at a specific time in history. Google Trends helps you to research the trends of numerous issues. These may include what the inhabitants of a specific country are searching for right now, or even the collective list of people's dreams all over the world.

Trendspotter sites are devoted solely to what might be coming soon. Try out trendwatching.com, meetglimpse.com or explodingtopics.com. Most have a free regular newsletter full of the latest data on their topic of interest. A search for Trend Analysis Tools Software will uncover many sites that provide you with the software that can help with your own analysis. Most require a subscription for the download.

Follow some futurists. They devote their lives to determining future possibilities and trends. Catherine Ball, Ross Dawson, Genevieve Bell, Gerd Leonhard, Soheil Inayatullah, Peter Diamandis, Tim Longhurst, Michael McQueen and Ray Kurzweil are just a few of the notable futurists worth following. A site called Compass, administered by the Association of Professional Futurists, provides rich and well-researched material on processes such as foresight and trend analysis.[17]

Search for trends in your everyday life. Look in shop windows for designs that are pushing acceptable limits. Watch for the response in others when they see a specific event or display. Note the way that large crowds show interest in a street performer. Attend a trade show, and observe which stands attract the most people. Take photos of everything and anything (with respect to other people), and openly look for opportunities to capture the image in an unusual way. Engaging in photography encourages you to look around for new possibilities. Travel to new places can stimulate you into looking for different possibilities.

SuperForecasting

A group of people called superforecasters exhaustively research trends and other relevant data. They are individuals who are more accurate with their predictions than experts or the general public, and this has been substantiated by statistical measurement. The concept of Superforecasters was introduced by US academic Philip Tetlock, who spent nearly two decades collating and testing thousands of forecasts from so-called experts.[18] One of his observations: they were not very impressive.

In 2011, piqued by the IARPA (Intelligence Advanced Research Projects Activity) search for effective methods to forecast geopolitical events, Tetlock co-founded the Good Judgement project.[19] He received online nominations from the public, provided advice on effective forecasting, and measured the accuracy of people's predictions. The best forecasters achieved around 30% better than intelligence community analysts. The Good Judgement project won the IARPA competition in 2015.

Tetlock also maintained that aggregate forecasts were more accurate than individual ones. The concept of 'the wisdom of crowds' was first identified by Francis Galton in 1907, and later was popularised by James Surowiecki in his book "The Wisdom of Crowds". The book quotes an example from Galton in which he observed an English county fair contest to estimate the weight of an ox. Nearly 800 people arrived at a median guess of 1,197 pounds - just one pound short of the correct answer.[20]

Initiated in the UK, the Cosmic Bazaar is an online tournament inspired by the IARPA project and the Good Judgement program.[21] Around 10 000 forecasts have been made by over a thousand forecasters from 41 government departments since 2020. There are between 30 and 40 questions live at any one time. The users are later ranked according to the validity of their predictions. Governments want to enhance the predictive abilities of their own employees, given that this might provide insight into the future actions of other unstable global leaders.

So what do these individuals do to achieve their results? According to Tetlock and Gardner in 'Superforecasting: The Art and Science of Prediction', they exhibit a range of compelling characteristics, including open-mindedness, political knowledge, and cognitive ability. They stay open to opposing perspectives, revise their beliefs often, think in terms of uncertainty, aim to be humble, use precise questioning, and detect patterns.[22]

Humility is another key virtue that good forecasters display, and this is demonstrated in many ways. They include: staying open to data that you personally don't agree with; accepting when your prediction was way off the mark; and immediately clarifying what you might do to improve your forecasting the next time around. And especially, you would avoid making rash public statements about future events that you claim are 100% likely to occur. 100% means that it will happen, without exception. Only a fool – or a very arrogant person – would announce a guaranteed prediction.

Predictive Exercises to Expand Your Predictive Ability

- Develop your 'hindcasting' skills. Write down a prediction; after the event, check to see how close you were to the outcome, and then note what you could do more effectively the next time you make a better prediction.

- Explore basic AI and machine learning tools that can help with predictions. Start with simple models available online and use them to predict outcomes in areas like consumer behaviour or financial markets. Reflect on the difference between machine-based predictions and human intuition.

- Clarify the difference between possibilities and probabilities in your own life. Name one of each for tomorrow.

- Nominate one specific industry or profession. Now search for three trend analysts in that field.

- Choose a current event or trend and create multiple future scenarios based on different variables. For instance, predict the future of remote work in the next decade considering various factors like technology advancements, societal changes, and economic conditions.

- Improve your research skills eg how to effectively use a large language model such as ChatGPT. Then use those skills by making a prediction about a specific issue eg the results of your favourite sporting team.

- If you are a gamer, play computer games that place you in scenarios about future events eg how to engage in international business transactions.

- Practise developing the core question for what you want to predict. This question would generally refer to the specific date and circumstances.

- Follow a successful trendsetter online for several months, and take note of the changes in their predictions.

- Talk generally of the future in terms of probabilities. Rarely make a definitive Yes or No prediction. That can even indicate arrogance. Instead, you might determine that "there is a 70% chance that the meeting on Friday will be cancelled."

CHAPTER 3 IN SUMMARY

- The chapter emphasizes the importance of distinguishing between possibility (the mere potential for an event to occur) and probability (the statistical likelihood of it happening), advocating for the use of percentage probabilities over Yes/No predictions.

- Modern predictive analytics tools, powered by Big Data and machine learning, are crucial for calculating probabilities in various domains, from weather forecasting to business strategy, though they are limited by the quality and relevance of the historical data they rely on.

- Introduced as a foundational concept, Bayes' Theorem is explained as a method for updating predictions based on new evidence, illustrating how prior knowledge and new data interact to refine probability calculations.

- The chapter discusses the importance of trend analysis and the influence of social media and other platforms in shaping and predicting trends, emphasizing the utility of both observational methods and specialized software.

- Highlighting the concept of "superforecasters," individuals who consistently outperform others in prediction accuracy, the chapter underscores the value of humility, open-mindedness, and constant revision in improving one's predictive abilities.

CHAPTER 4: PRACTICAL PREDICTION

Life can only be understood backwards;
but it must be lived forwards.

SOREN KIERKEGAARD

Did The Simpsons Really Predict the Future?

You may love or hate the Simpsons, but you have to be intrigued by the predictive efforts of the scriptwriters for this ongoing TV cartoon series. At least 30 different events are claimed to have been forecast in earlier shows. They included the US presidency of Donald Trump (first alluded to in 2000), the tragic loss of the OceanGate submersible near the Titanic (17 years before it happened), Lady Gaga flying into the Super Bowl stadium while hanging from cables (on the show in 2012, 5 years before it occurred) and the revealing by Homer of the equation for the mass of the Higgs boson particle (14 years before it was revealed in 2012).[1]

There are several explanations for these predictions. Given the satirical nature of their shows, the writers possibly never intended to create predictions. They just wanted to be ridiculous. An adjoining point is that if you make hundreds of predictions, some of them are probably going to come true, simply by chance. Take note that little media attention is given to the many predictions that did not eventuate. Perhaps the most valid explanation is that the script-writers combined their wacky sense of humour with some diligent research, and continually delighted in the public's exaggeration of their purported predicting.

Your family life (hopefully) does not directly reflect what happens to the Simpsons, although the predictive element in the TV series might be worth emulating. All parents want their children to grow up as well-balanced healthy young people, and so they visualise who their children will be in the years up ahead. They might then backward map through to now, and determine what support their young ones will need to become inspiring grown-ups. So how can we know what best prepares children for engaging effectively in adult life?

Predicting A Child's Future

Longitudinal studies offer powerful predictive indicators. These studies are conducted with nominated children, often over decades, and reveal how childhood experiences and behaviours can influence everyday functioning when they become adults. The UK-based 7UP series popularized this approach with their long-term focus every seven years on the lives of fourteen children. 63 Up was broadcast by ITV in 2019.[2]

The Dunedin Multidisciplinary Health and Development Study[3] is a landmark program that has researched the lives of 1037 New Zealand children who were born between April 1972 and March 1973. One of the Dunedin long-term findings is that study members who exhibited better childhood self-control tended to have younger-looking brains in adulthood. The study also found that experiencing smoking, mental health problems, and obesity during adolescence made it more likely that the brain would show signs of faster aging in adulthood. A quality childhood really counts.

Effective parents and educators develop resilience in children so they will thrive through anything that happens up ahead. The secret is to implement protective factors. These are interventions in children's lives that can lead to healthy development, and might involve friendships, mentors, and engagement in sport and recreational activities.

On the opposite side, risk factors can compromise child development. Being allowed to roam around the local streets at midnight when they are ten years old is one example. They also include experiences that cause toxic stress and anxiety. The world will be an even better place in the future if we lower the risk factors and raise the protective factors in children's lives today.[4]

Analysis of everyday data helps us to partly predict a child's future. With predictive policing, we can calculate with up to a 90 per cent certainty the likelihood of some teenagers committing an offence, especially if there are many risk factors in their lives.[4] Medical analysts can assess with up to an 80 per cent probability whether children will contract the flu, eight days before they feel ill. The GPS on the child's mobile phone can signal where they were today. If anyone in the child's group indicated on social media that they were feeling sick at the time, your child's chances of contracting that illness are much higher, given their physical proximity to the unwell person.[5]

Ensuring that children can read competently is a valuable protective factor. When they are highly literate, young people tend to be more successful in other education subject areas. One effective strategy for boosting literacy is to occasionally ask them to predict what might happen next in the book they are reading. Another variation is to show them the title and the cover of the book and ask them to conjecture on the inside content. While sharing a book with early-stage readers, you might pause before turning the page and ask: So, what do you think will happen next? These practices encourage a child to build connections with their prior knowledge, and to develop hypotheses about the intent of the author.[6]

A further protective factor is to show children how to set and achieve goals. This goal-setting can begin with 4- or 5-year-olds, and might be initiated with a simple conversation about what they would like to do this weekend. Once children reach 7 or 8 years of age, most have the brain capacity to think abstractly about near-future events. There is merit in setting goals that are just out of their reach. Building their goals into a plan for the whole family will provide ongoing support and a deep sense of inclusivity.

Even writing goals down will give children a greater chance of making them happen. You might ask your child a series of questions such as: What do you want to achieve? How and when will you do it? Goalsetting generates a belief that they have some control over predicting and creating their own future. When times get tough – which they invariably will – goalsetters are more likely to resolve the issue, rather than wait for life to dictate to them.

Global education systems today focus on what we call capabilities for children. A capability is something that helps them to become more capable of coping with an uncertain world up ahead. The list can include literacy, numeracy, initiative, work agility, resilience and empathy. My advice is to include predictive capability on that list. Given the exponentiality of change in their future, children's predictive acumen will stand them in good stead. This is why we need to stop barraging children with the worst excesses on the planet. If they become convinced that the future is dystopian, their predictions will reflect this perspective.

Science Fiction Prediction

Science fiction has a long history of predicting future technologies and lifestyles. The sci-fi genre is attractive to us in part because we are fascinated by what lies up ahead. Some movies turn out to be accurate, while many do not. Even amongst the so-called accurate ones, some would have eventually occurred simply through chance. One concern is the dystopian bias that is often built into sci-fi script in order to sensationalise the action. Spoiler alert: the future will not always feature dramatic and terrifying circumstances. It will also include plenty of routine everyday events as well.

Some sci-fi writers do manage to prophesise or influence the future. Science fiction can become science fact in a short period of time. In 1998, a movie called 'The Truman Show' featured Jim Carrey, who acted out the part of Truman Burbank. This character was unknowingly followed in everyday life by the OmniCam Corporation, who portrayed Truman's daily adventures in a high rating TV show. If you have since enjoyed *Survivor* or *Big Brother*, you can probably thank Truman.

What else did sci-fi predict? The Dick Tracy comic strip in 1946 featured a two-way wrist watch. At the time, it seemed far-fetched, and yet users can now make calls on their smart watch. Even late in the 20th century, international telephone voice calls were charged by the minute. Now, video calls on WeChat, Facetime and other apps are free. This form of video technology first featured in a 1911 novel called *"Ralph 124C 41+"* by Hugo Gernsback.

Star Trek provided some compelling indications about future technologies, with references to voice activated computers, debit cards, and audio translation back in 1979.[7] George Orwell wrote his dystopian *'1984'* back in 1949, and focused on the consequences of mass surveillance and totalitarian regimes. The technology included 'telescreens' for observation, and 'SpeakWrites', which could convert speech into text. The concept of 'Big Brother' still applies in real-life TV shows, and in the minds of some power-hungry leaders.[8]

The "hunter seeker" assassin drone in Frank Herbert's 1956 novel *Dune* was an indicator of the modern-day drone. *Star Wars* also featured small autonomous flying devices. The first commercial drone permit was not issued until 2006 by the U.S. Federal Aviation Administration. Further back in

1936, Raymond Z. Gallun predicted robo-bees that could pollinate crops. This has been popularized in the series *Black Mirror*. In a 1964 New York Times article, Isaac Asimov first wrote about cars that would have a 'robot brain', and could be set to drive to a particular destination by itself.[9] Although the required technology is more complex than originally thought, autonomous cars will become commonplace in the 2030s.

Other movies and books were not so successful in their predictions. Many of the earlier sci-fi efforts featured flying cars and old-fashioned landline phones. Hoverboards are still to make an appearance. The post-apocalyptic *Mad Max*, in which looters featured in a world without fuel, has thankfully not yet come to pass. So too the *Predator* series, in which aliens devastate the planet.[9] Two questions need to be asked about movies like this. Why are the aliens always featured as the bad guys? And: Why do so many of these movies focus on destroying New York? Please continue to enjoy the sci-fi genre, while also maintaining a neutral stance on the negative biases the script may contain.

The Sporting Advantage

Science fiction might encourage you to predict the far future, although sport encourages you to forecast what will happen in the next split-second. Moving early to the ball or opponent gives you a distinct advantage in most sports. You will have more time to balance correctly, and to deliver the best possible play. Predicting your opponent's tennis shot before he or she hits it gives you an advantage of perhaps one second or more. Researchers at QUT (Queensland University of Technology) developed an algorithm that trains you to predict shots from the other side of the tennis net, thus giving you that precious advantage.

The Semi Supervised Generative Adversarial Network architecture analysed the data from thousands of shots played by Novak Djokovic, Roger Federer and Raphael Nadal at an Australian Open. The context for the shots – whether it was played in a low pressure first round, or during a match point in the final – was added to the data. The algorithm copied the thinking of these top players who were constantly predicting their opponent's next shot. This architecture can be employed in virtual reality games, and in real life with other players. Interestingly, Federer was harder to predict, given his prodigious ability to re-adjust his stroke play at the last split-second.[10]

World class players in any sport appear to have more time. It is partly because they have predicted what other players are going to do, sometimes even before the opposition players themselves know what they are going to do. The GOATs (Greatest Of All Time) seem to possess a sixth sense about their opposition's movements. In team sports, the exceptional players can view the entire field, and calculate the sum total of the movements of all players at once. Football goalies use every possible predictive indicator when an opposition player is shooting for a penalty goal.[11] Eye movements, twitching of muscles, and excessive shuffling of the feet might all give some small hint of what shot will be kicked. Some goalies claim they can read minds.

How can you practise this predictive skill? It eventually needs to work for you in the heat of the sporting battle, although one key activity helps with this. When you watch a sport broadcast, call out the likely plays in the next few seconds. This is different to simply watching the event in real time, and commenting on what is happening (often accompanied by some unkind opinions about the referee). The same process can be used for movies you watch. Occasionally, predict what is about to happen next. Please only practise this predictive activity when your long-suffering partner or flatmates cannot hear you.

You may not consider chess a sport if you are more attracted to physical pursuits, although it is just as much a competitive challenge. A casual game might last for an hour, although championship games can progress for six hours or more. Everyday players might think one to three moves ahead, while Grandmasters can plan ahead by twelve moves or more. These master players are also assessing which play their opponent might make in the next four or five moves. In their own way, they are reading the entire field – the board full of chess pieces – and predicting what lies up ahead.[12] Given this predictive ability, chess is a powerful analogy for the sporting and corporate world.

Insuring Your Future

Most industries will not thrive unless they engage in predictive analysis. One specific corporate group would be bankrupt very quickly if it did not forecast the future. The insurance business is based upon probabilities, and they certainly do not use a crystal ball to calculate the odds. If you have ever purchased insurance for a car, for a house, or even for your own life, some predictive analytical tools will have been used to assess what you will pay.

This amount will vary, depending on your insurance history, the amount and type of coverage, and especially, your age. The actuaries who calculate your life insurance payments develop a life table which combines mortality rates with relevant personal data and possible future trends to determine your life expectancy.

Your age makes an obvious difference. If you are 70, you are closer to passing away on average than a 20-year-old, and this will be reflected in your life insurance premium. Car insurance will be the opposite, with the 20-year-old considered a greater risk because of the lack of experience. And the heightened testosterone, in too many cases. Other personal information such as your credit history, any prior claims, your marital status, your smoking history, and where you live, will influence the amount you pay.[13] All of these indicators are included on a complex actuarial table, and you will be placed somewhere on that chart.

The majority of insurance companies worldwide have a net profit margin of 2 to 3%.[14] However, depending upon the specific focus - property, personal goods, health, life - the margin may be 15% or more. The top 20 insurance companies globally have a combined market value of nearly $1.6 trillion, with assets of more than $11.2 trillion.[15] These companies ensure that they will receive more in the premiums paid than any claims that are made. As with most businesses, if they set the premiums too high, then fewer will buy the insurance. If they set them too low, they may go out of business. One significant factor will influence their profitability in the future. That factor is climate change.

The next time someone claims that climate change is a myth, and is rarely supported by hard-working business executives, ask them what insurance company CEOs would say about the issue. Insurance is one industry that takes the changing climate very seriously. If some of the long-range forecasts are accurate, these companies will incur massive payouts. As a result, insurance premiums for property are likely to dramatically increase for most people in future years. These exorbitant premiums may lead to fewer people declining on their insurance, which possibly means that the fees will become even higher for those who maintain their cover. Some climate change risks have already become uninsurable.

Your Future in Your Genes

DNA testing, also called genetic testing, is one way you can partially predict your health and lifespan up ahead. These tests can identify mutations in your genes, chromosomes, or proteins. Explicit predictions cannot be given about time of death, although the tests are a reliable method for discovering mutations that may increase your risk of cancer.[16] Once you have received the results, you may choose behaviours or treatment for lowering your future risks. Insurance companies may ask for any DNA test results when they are calculating your life insurance premiums.[17]

In the Icelandic capital of Reykjavik, a company called deCODE genetics has sequenced over 400,000 whole genomes, more than anyone else in the world. These large numbers allow scientists to determine the inherited risk of chronic illnesses such as schizophrenia, Alzheimer's, and various forms of cancer. Further research is being conducted into the connection between our genomes and other aspects of your life. These can include food preferences, your personality, and your capacity to build friendships. More scientists are now asking: Is your behaviour solely your own choice, or is it pre-determined by your own biology?[18]

Even if your genetics play a part in your life choices, they can only influence your disposition towards specific activities. You still need to make the choice to engage in those activities. Your DNA is not your destiny, although it can flavour your choices. DNA tests provide clear data about the genes that constitute who you are. The tests can clarify if you have a stronger risk of developing specific conditions.

In one British study, the scientists conjectured that you may have a form of 'hobby gene', which predisposes you towards a specific hobby.[16] There may also be a form of 'talent gene', although you will still need to do the training or practice necessary to improve. While DNA is important, the overall environment plays the key role in determining how closely you follow your genetic dispositions.

There are different predictive indicators in your everyday life. Researchers describe an 'inherited affect'[19] that can be passed down through time. When children experience unsettling physical or emotional abuse, the resultant effects are not just felt in the adult lives of those children. Intergenerational trauma can impact the lives of their generations far ahead. The study of epigenetics clarifies how lived traumatic experiences can be passed down through their children's children. The effects can include DNA modifications, abusive cultural messages, persistent aggressions, and normalization of hatred and cruelty.[20]

Toxicity can also be inherited. Washington State University research with rats indicated that male infertility could result from exposure to a pesticide called vinclozolin. The issue is that the infertility did not evolve from exposure for the infertile rat, but to its great grandmother three generations previously.[21] The implications for the human race are disquieting. Although the research has not been conclusive, there are strong indications that human male sperm count globally may have declined by up to 50% in the past five decades. Exposure to environmental chemicals is one possible reason.

Can you inherit a psychological affliction such as anxiety? The probability of you developing the disorder is between two and six times higher than normal if a family member or close relative suffers from anxiety.[22] This does not mean that you are guaranteed to experience the same disorder, although the likelihood increases. Varying life circumstances matter more at the personal level. An 8-year-old's fear of being left alone is very different to an older person's concern about ill-health.

However, a child's circumstances - such as being abused or bullied - means that he or she is more likely to suffer from anxiety in later life.[23] Children observe their parent's behaviour in stressful situations, and may learn to respond anxiously in the same way. Authoritarian, over-protective, or even neglected parenting can lead to higher instances of anxiety in children. Being a parent is probably the most significant influence you can ever have on the future lives of other people.

The Ripple Effect

The 1998 movie *Sliding Doors* focused on the dramatic changes that can occur in your life from a single choice.[24] Set in two parts, the movie featured a character who rushed to catch a train in the first part. When she arrived home, she found her boyfriend in bed with another woman. In the second part, she missed the train and waited patiently for the next one a few minutes later.

Upon arriving home, she found her boyfriend by himself. The visiting woman had already left. Highly divergent consequences resulted from which train she caught.

According to chaos theory, small actions can have significant consequences. Essentially, you can instigate a vast array of future events merely because of some actions that you take today. The world is a complex place, and anywhere between one and a billion factors can intervene to create a future event. Historians conjecture that the catalyst for the start of World War 1 was the assassination of the Austrian Archduke Franz Ferdinand on the 28th July 1914.[25] This led to other political events such as Germany's support for Austria invading Serbia, which contributed to the start of the war.

Scientists refer to the butterfly or ripple effect when considering the causation of weather conditions. The flap of a butterfly's wings reputedly leads to a tornado developing halfway around the world. Is it possible to manage these ripple effects in everyday life, given how many other factors will come into play? With some actions, it is conceivable. Giving your child a hug before she leaves for school might lead to her receiving a community award three months later for her social contribution. Buoyed by your morning affection, she becomes involved that day in a school club that gives support to struggling families. Later on, she is acknowledged for her work.

The endless ripples from a specific action can create a nightmare for organisations. The World Health Organisation (WHO) discovered this when they sprayed malaria-plagued Borneo with DDT in the 1950s. While this worked to eliminate the mosquitoes, it created consequences for many villagers' thatch-roofed huts. The spraying killed the wasps that ate thatch-eating caterpillars, which led to many roofs collapsing. The DDT moved through the food chain, with insects, lizards, and even cats being affected. The rat population increased as the cats died out, which led to an outbreak of typhus and sylvatic plague. The solution? The WHO parachuted live cats into Borneo.[26]

I can attest to the power of these ripples. Back in 2000, I wrote a book called *The Ripple Effect*,[27] which ended up selling in its thousands. In 2019, an Australian woman who had read the book recommended me to a work colleague in the UK. The UK woman contacted me to arrange for a presentation with their staff in Australia in July of that year. After a successful day with their group, I was then invited to present at a conference in London in January 2020. On the 1st night I arrived there, I met a beautiful woman called Lana, who worked with the same UK company. Two years later, we married and have now settled happily in Brisbane. The delicious quirk is that The Ripple Effect itself caused a ripple that profoundly changed both of our lives.

Sometimes you just need to trust to the Universe. Not everything can be rigidly controlled, and the serendipitous moment can occur on a billion to one chance. I even had one such event take place while I was writing this book. Back in Chapter 3, I listed the names of some exciting global futurists who are worth following.

Just as I finished writing the paragraph, I received a call from Sydney-based Tim Longhurst, a professional colleague and futurist who I had not seen or heard from for over ten years. He had been attempting to reach the father of another Tony Ryan, and had called me by mistake. Suffice to say, his name was (rightfully) already included on that list of quality futurists. The chances of him calling me after ten years at that precise moment are beyond calculation.

Predictive Exercises to Expand Your Predictive Ability

- Practice making small, immediate predictions in everyday life. For example, predict how long it will take to commute to work, the outcome of a conversation, or the weather for the next hour.

- When you watch TV sport, call out the likely plays in the next few seconds. Don't do this when your partner is also watching.

- Use a similar process when viewing a new movie. As you watch the action, predict some possible future scenarios, based upon what you are seeing.

- When reading a story to a child, pause before turning the page and ask them to predict what will happen next. This exercise helps develop their critical thinking and predictive skills.

- Set up a daily self-prediction journal. Write it the night before. This can help you become more aware of the factors influencing your daily life and improves your ability to foresee short-term events. By regularly reflecting on your predictions, you can identify patterns and improve your predictive accuracy.

- Arrange a DNA test, and discuss the results with your doctor. The test cannot predict your date of death, although it may give strong indications for you to change behaviours or eating habits.

- Arrange a quote for some life insurance. It will indicate to you how long the insurance company expects you to live.

- While the ripple effects from your actions can be complex to predict, make a habit of occasionally forecasting at least one consequence from an everyday action that you take.

CHAPTER 4 IN SUMMARY

- Long-term research, like the Dunedin Multidisciplinary Health and Development Study, demonstrates how childhood experiences can significantly influence adult life, particularly through the development of self-control and resilience.

- While some sci-fi predictions have come true, such as smartwatches and video calls, others are exaggerated for dramatic effect. The genre reflects our fascination with the future but too often portrays a dystopian bias.

- Elite athletes and chess grandmasters gain a competitive edge by anticipating their opponents' moves. This predictive ability is developed through experience and can be practiced by forecasting outcomes in everyday activities.

- The insurance industry heavily relies on predictive analysis to set premiums based on risk factors, with climate change emerging as a critical future concern that could drive up costs and make some risks uninsurable.

- DNA testing can indicate health risks, while studies in epigenetics reveal how trauma and environmental factors can affect future generations, emphasizing the complex interplay between genetics and environment in shaping behaviour and health.

CHAPTER 5: PREDICTION BY MOTHER NATURE

Any astronomer can predict with absolute accuracy just where every star in the universe will be at 11.30 tonight. He can make no such prediction about his teenage daughter.

JAMES TRUSLOW ADAMS

The Annual Circle of Life

You can set your annual calendar by the timing of migratory bird patterns. One such example is the American cliff swallows of Capistrano in Orange County, California. Around October 23rd each year, they migrate nearly 10,000 kilometres away to the southern summer in Goya, Argentina. Their return on or about March 19th - St Joseph's Day - is a celebratory tourist attraction in Capistrano.[1] Partly due to increased construction activities in the town, fewer swallows have been returning in more recent years.

When we talk about world record distances, none can compete with Arctic terns. Sometimes referred to as the champion of migration, they fly between the Arctic and the Antarctic each year, a one-way distance of at least 19,000 kilometres. This allows them to enjoy two summers, and to experience more daylight than any other creature on earth. In an average lifespan of thirty years, they can end up flying more than 2.4 million kilometres.[2]

With their flight paths, it is believed that birds determine the correct direction to fly by detecting the magnetic field generated by the Earth's molten core. The light from the sun can also play a part in finetuning their direction. Along with hibernation, migration is a core annual habit of many animals. So what triggers these events? Indicators such as a change in food supplies, the amount of daily sunlight, and lowered temperatures will signal that the migration or hibernation is near.

Migratory and hibernation patterns are part of the awe-inspiring rituals that occur during our trip around the sun each year. In earlier times, knowledge about the near-future of nature's patterns helped humans to plan their everyday existence. People became expert at hindcasting. Year after year, they would observe the changes of the seasons, and then add to their knowledge base of what to expect at different times of the year. Their powers of observation helped to accurately predict what was going to happen next year. Knowing which animals would return to the grasslands, or when to expect the rains, determined their very survival.

What can we learn from nature? Everyday predicting can be done with two core methods. One is to calculate the linear progression of events over weeks or months, such as improving your wellbeing and energy levels with regular nutritional meals. If no other unusual circumstances intervene, forecasting with linearity is feasible. The second method is to gain knowledge of events taking place in a cyclic fashion, such as the 24-hour daily occurrences or the 12-month yearly cycle. When a regular event occurs the following day or year, you recognise it as a predictable part of your existence.

All your life is indeed a circle. You wake up, have breakfast, go to work, and continue through the day until you wake up the next morning. Over 90% of what you do today will be similar to what you accomplished yesterday, or the same day a week ago. For most people, it is relatively easy to predict what they will be doing tomorrow, given that it is so similar to what was done today.[3]

Research by the Media Lab at MIT (Massachusetts Institute of Technology) monitored people going about their everyday lives while wearing small tracking devices. This monitoring recorded their voice tones, their body language, and where they travelled. The researchers concluded that humans are indeed creatures of habit.[4]

Your predictive brain is aware of these patterns that make your life more stable. The foods you consume at different times of the day and the year, the clothing you choose to buy and wear, the sport and social activities in which you engage, are all part of this predictable pattern. It provides stability to know what is likely to happen up ahead. If something like an anniversary occurs at the same time each year, it is usually to your advantage to remember it.

Profitable business cycles require a predictive element. Clothing department stores follow seasonal patterns, while food outlets adhere to daily and seasonal cycles according to a community's needs at the time (eg hot soup on a cold day). Ice cream sales are between two and four times higher in Summer. This is partly because some ice cream stores close during the winter months, due to a lack of demand. Umbrella sales increase during a rainy season, and even on a day-to-day basis depending upon sudden downpours. Predictive analysis of future profits in many businesses is based upon daily, weekly, monthly, quarterly, and annual cycles.

Your Body Cycles

Have you ever heard that all of the cells in your body are replaced every seven years? This is not even close to the reality. Different cells regrow or develop at widely different amounts of time. The cells in your stomach lining can renew as often as every two days, while skin cells are replaced every two to three weeks. White blood cells last from a few days to a week, and red blood cells survive for about four months. Your fat cells exist for up to ten years. Brain cells do not regenerate, although the cells in the hippocampus, which is responsible for memory, can regrow with worthwhile stimulation. Tooth enamel and eye lens are never replaced.[5]

Your circadian rhythm, which follows a 24-hour cycle, creates the physical and mental changes you make through the day in response to light and dark. If you wanted to calculate your natural circadian rhythm, you would turn off your alarm clock and take note of when you wake up and when you perceive you are hungry.

Have you noted when you feel more energetic? For most people, their high point is late morning through to lunchtime. Low points are usually around 2 to 3 pm, and often from 2 am to 4 am in the early morning. Predicting these ahead of time can help you to plan your day more effectively.

Many other predictive cycles are part of everyday life. Between the ages of approximately 12 and 53, women will experience up to 480 periods. They will spend nearly ten years of their life on their period. The average age of an initial period in the 1800s was 17, whereas girls today usually begin around the age of 12.[6] Some men – around a quarter – are believed to experience regular hormonal changes. These are rarely as prevalent as those that are experienced by women. The most obvious one is usually with their testosterone levels, which decline by about 1% a year after the age of 40.[7]

It's In the Stars

Regular movements in the celestial bodies are highly predictable. Because of their dramatic visual impact, lunar and solar eclipses have engendered both awe and fear in people for thousands of years. These eclipses can be accurately predicted up to one thousand years ahead. Beyond that, the calculations are compromised by the slight rocking and wobbling of the moon as it moves through the sky.

The Saros cycle - 18 years in total - outlines the length of time it takes for the Sun-Earth-Moon system to return to the same triangular configuration. If you witnessed a total solar eclipse in 2032, there would be another one 18 months later, although in a different part of the Earth. It would take approximately 375 years before a similar eclipse would occur in the same place.[8]

This predictive analysis has special merit when we investigate NEOs (Near-Earth Objects), chiefly asteroids and comets. Even a small asteroid one metre wide at impact with the Earth would be highly destructive. Expand that to ten kilometres in diameter, travelling at least 30 kilometres per second, and life would probably cease to exist on the planet. As well as the damage caused by the direct impact, the resulting dust cloud would blanket the earth for years and create a nuclear winter. If it lands in the ocean, the resulting tsunami would be monumental.

An asteroid with a diameter of 300 metres might impact with the Earth once every 70,000 years. A 600-metre asteroid may collide with us every 200,000 years. No large NEOs are thankfully predicted to hit in the next hundred years.[9] By then, advanced technology will hopefully be able to divert their path.

What else might affect us from outer space? Solar cycles are one such example. The charged gases in the Sun create an electrical field, which causes its North and South poles to completely flip. Eleven years later, they upend again. Solar flares increase at specific stages during this solar cycle, which can affect radio communications on earth. Those same flares also create more prominent aurora. If you intend to holiday near the North or South poles on earth, and hope to witness an aurora, choose the time that correlates with these solar flares.[10]

Could the ancients predict such astronomical events? The records indicate that over 2000 years ago, these early astronomers could accurately determine the timing of an eclipse. They spent just as much time searching the sky as many of us do with our computer screens today. In the 18th century, Edmund Halley explored historical records for comet sightings, and in combination with Newton's theory of universal gravitation, calculated the next appearance of Halley's Comet (named after him). He also predicted a total solar eclipse over London on May 3rd, 1715. This was accurate to the nearest four minutes, and within 20 miles.[8]

Weather Forecasting

In earlier times, humans didn't just look to the sky at night-time. The weather during the day had immediate significance for them as well. Over 2,500 years ago, the Babylonians observed clouds to predict the weather. Around 340 B.C., Aristotle wrote *Meteorologica*, a book that outlined many theories about the formation of wind, rain, clouds, thunder, and lightning. This text became the optimum weather reference for the next two millennia. The irony of weather forecasting in earlier times was that the weather arrived before the weather forecasts, given that the horse and its rider could not move as quickly as the clouds.[11]

Observing cloud patterns is still a reliable indicator today. A halo around the moon can indicate that poor weather may be imminent. This lunar halo more often happens during winter months, and is caused by light passing through the ice particles in high altitude cirrus or cirrostratus clouds.[12] When the sky is nearly full of the wispy cirrus, it is more likely to rain within the next day. Fluffy cumulus clouds generally indicate that fair weather will continue. The most dangerous clouds are the large cumulonimbus, which are sometimes referred to as 'thunder heads'. They clearly signal their intention with their dark bottoms, and are the most likely to lead to unsettling storms.[13]

Modern forecasting systems are partly based upon the release of radiosondes, a small box equipped with weather instruments and a radio transmitter that are carried by balloon to a height of 30 kilometres.[14] The combination of balloons, satellites and weather stations provide feedback to weather bureaus about the current state of the atmosphere. Their latest tools include numerical weather prediction models that can accurately determine the likelihood of various weather outcomes. The most important contribution is provided by meteorologists who can decipher the complex data and provide the public with accurate forecasts.[15]

The history of weather forecasting is littered with tales of animals predicting the weather. Ask anyone who spends time on a farm or in the bush, and they will likely regale you with endless myths about animals that can forecast a pending storm. In part, they are probably correct. Animals are highly aware of minute changes in barometric pressure and the humidity in the air.[16]

Some dogs and cats can hear the distant rumble of thunder and smell the metallic odour in the air that occurs just before a storm. Their behaviour often becomes erratic, and they may hide for safety. Cows reputedly lie down before a storm, but they do this at other times as well. Sharks can detect the decrease in water pressure before a hurricane and may swim into deeper water to escape the worst effects of the tempest.[17]

Birds sometimes fly lower before an impending storm to escape the stronger higher winds. Both birds and bees appear to return to their nests or hives. Frogs need water to lay their eggs, and reproduction is more likely after heavy rain. You can sometimes hear them croaking their mating calls before the rain begins.[18] Even some humans contend that they can predict a storm the day before, due to a headache caused by the drop in barometric pressure.

In 2022, the UN introduced a global plan for every person to be covered by early warning systems by 2027. With a price tag of just over $3 billion, these systems will develop models that forecast impending bad weather, prepare local populations ahead of time, and send out alerts to anyone who might be impacted. The strengthening of local expertise to guide these systems will be one of its most critical factors for success. Floods and droughts account for 75% of all climate disasters, and early alerts are especially critical for agricultural-based economies.[19]

Earthquake Forecasting

Because of the horrendous damage caused by earthquakes, seismologists constantly search for effective warning indicators. Groundwater changes, electrical changes, and foreshocks may offer some small chance of prediction. We know that quakes are more likely to happen at plate boundaries, or where they have occurred previously. When they are going to happen is still unknown. However, some creatures give indications that they can detect earthquakes beforehand.

In 2009, a scientific researcher in L'Aquila in Italy noticed that a colony of toads disappeared three days before a quake struck the region. They did not return to the site until ten days later.[20] Perhaps they can detect the gases and charged particles that are sometimes released prior to an earthquake. Because of the large rock pressures, other creatures may be able to sense the resultant ionization of the air with their fur or sensitive skin or exoskeleton. Anecdotal evidence indicates that many reptiles, insects, and animals demonstrate strange behaviours up to several weeks before a quake.[21]

Even with these vague indications, no scientifically valid process exists for reliably predicting a large quake. Even if we could do so, the prediction of such events raises a difficult dilemma for seismologists, and it is a concern for anyone who can capably predict future events. There is always the chance that they might not be correct, and the release of a public warning can create terror in a population. As well, the costs for immediate preparatory action can be significant. Imagine ten million people trying to leave a city all at the same time.

Some earthquake apps can provide some small measure of advance earthquake warning. The tiny accelerometers in smartphones can detect vibrations that indicate an earthquake is happening. The collective signals are then received by an earthquake detection server, and the results are immediately sent to other phones in surrounding locations. It may only be a few seconds of warning, although that can be enough time to dive underneath a table.[22]

Scientists have even been blamed for not predicting a quake before it occurs. Six seismologists were arrested for manslaughter after failing to predict the L'Aquila event in Italy and were sentenced to six years in prison.[23] They were eventually released after an extensive public outcry. This legal entanglement ensures that scientists will be overly cautious next time, and less likely to release a warning when it is more likely to be needed.

Climate Change Forecasting

Scientists must feel the same way when it comes to the quandary of climate change that is caused by human activity. The evidence is overwhelming that it is taking place, and yet climate change sceptics continue to manipulate the media and cast doubt in the public mind about the realities. All of the climate myths, such as the sun is warming (no, it's getting cooler), the IPCC is alarmist (their predictions are more likely to underestimate the climate response), have been disproved.[24]

Here is a reality check for all of us. Climate change is categorically occurring because of human activity, although a dramatic transition to renewables may initially cause even more harm than good. Yes, you read that correctly. Let's start with the climate change evidence.

The emission of greenhouse gases from human activity since the late 1800s have led to a global temperature rise of 1.1 degrees C by 2021. The warming over land masses is higher than the global average, and more than twice as high in the Arctic.[25] If we average this rate of increase into the future, we will reach at least 1.5 degrees C by 2050, and between 2 and 4 degrees C by 2100. Those latter figures are variable and are predominantly dependent on our efforts to rein in global emissions. At the time of writing, the world recorded its hottest day ever on 21st July 2024. The average temperature was 17.09 °C (62.76 °F).[26] One swallow might not make a summer, but billions of contributory effects will certainly create a hotter one.

The predicted climate change consequences include increasing heatwaves, longer warm seasons, shorter cold seasons, affected rainfall patterns, sea level rise, and increased melting of glaciers and ice sheets. Low-income people in poorer countries will be affected the most. Their livelihoods will be more heavily impacted, especially if they live close to sea level. They also spend a larger share of their income on food, which is likely to considerably increase in price.

This mega-issue will not be easily resolved. A rapid transition to renewable energy may initially be even worse for us. The 46,000 coal, oil, gas and nuclear powered stations at present would need to be replaced by over 580,000 wind, solar and hydrogen stations. This rapid transition would result in even more fossil fuel consumption needed for the construction of those power stations, as well as for the many electric cars, solar panels, and wind turbines.

Australian-born geologist Simon Michaux claims that "Since 400 BCE, various civilizations dug up 700 million tonnes of metals (everything from bronze to uranium) prior to 2020. But a so-called green transition will require mining another 700 million metric tonnes by 2040 alone."[27]

Future Solutions

All is not yet lost. In just a few decades, we have predominantly resolved issues such as lead, CFCs, asbestos, and tobacco. Our collective capacity to create tangible environmental improvements is clearly in evidence. Africa's Great Green Wall is one such example. Initiated in 2007, this massive project involves planting trees along the entire width of the continent from Senegal to Djibouti. It is 16 kilometres wide, and 7000 km long. While tree-planting still features strongly, the focus is now placed more on sustainable land use, job creation, and especially, peacebuilding.[28]

The Ocean Clean-up is another inspiring project. The stated objective is to clean up 90% of all plastic floating in the world's oceans. The Ocean Clean-up workers are presently focusing on the Great Pacific Garbage Patch between California and Hawaii, which contains over 100,000,000 kilograms of plastic.[29] If you access the comprehensive updates on 'Future Crunch'[30] or 'Fix The News'[31] or 'Good News'[32], you will discover many other heartening environmental projects.

There are three overarching options for contending with runaway climate change. The first is to dramatically lower our greenhouse gas emissions. One powerful process to support this is the concept of a circular economy, which is based on building simple products that can be cheaply and easily recycled. As part of this, the five 'R's (Refuse, Reduce, Reuse, Repurpose, Recycle) would become core to every aspect of our lives.[33] While most people make some effort with all of these, the political implications of this first option are significant. Hundreds of millions of workers would need to transition to other employment.

The second option is to rapidly deploy some highly unusual possibilities. Clean energy from nuclear fusion might be a gamechanger, although it is unlikely to be commonplace before the 2060s. Geo-engineering involves some mind-bending proposals, including solar radiation management and carbon dioxide removal.[34] These are not small projects. Imagine spraying seawater kilometres into the air to seed the formation of clouds that then deflect sunlight, or installing giant mirrors in space for the same purpose. While such a project may work for some parts of the planet, others might be hugely compromised. We would play with Mother Nature at our peril.

Asteroid mining may have more merit, although it will not happen any time soon. The lure is that some asteroids in our solar system contain unimaginable wealth. NASA's Psyche mission, set to arrive in 2026 between Mars and Jupiter, will explore this 226-kilometre-wide chunk of rock. If the asteroid's mineral wealth is confirmed, it would be worth more than the global economy at $10 quintillion. Another asteroid called Davida is estimated to be worth $27 quintillion.[35] The obvious issues are the large-scale mining in outer space, and the transport of the raw material to Earth.

The third option is the most likely, and probably the most politically palatable. Whether it is the most effective solution long-term is another matter. The option? To steadily implement Net Zero initiatives as rapidly as possible, without destroying the world economy in the process. Cities, rather than whole countries, would most likely become the catalyst for enacting effective projects. Over 700 cities in 53 countries already have committed to halving their emissions by 2030, and reaching net zero by 2050.[36] Copenhagen citizens have built ecologically stable practices into everyday life. Bikes outnumber cars by 5 to 1. Less than 30% of the population own a car. Nearly half of all daily commutes are done by bike.[37]

Steadily, everyone everywhere could contribute in similar fashion. The question is: Will they? Most people agree that change is necessary. The abject reality is that their answer edges into the negative when asked if they are personally prepared to contribute to that change. Please remember: the word 'everyone' means 'every one' of us.

Predictive Exercises to Expand Your Predictive Ability

- Think through the habitual patterns in your daily life. Predict at least five everyday actions (and the time it will happen) over the next week.

- Use cloud patterns to predict the next day's weather. Record observations of different cloud types (e.g., cirrus, cumulus, cumulonimbus) and their corresponding weather outcomes.

- Don't just rely on technology to find out about weather forecasts. Look for changes in animal behaviour eg the croaking of frogs before a storm.

- Track your daily energy levels, meal times, and sleep patterns for a month. Note any regular cycles and how external factors (like light exposure) influence them. Enhance awareness of your biological rhythms and predict optimal times for various activities.

- Ponder the daily, monthly, and annual patterns in the behaviour of people you closely know. Many of those patterns are easily predictable eg what coffee people drink at the same time each day; the places they want to visit during summer.

- Stay aware of seasonal changes in weather and animal behaviour eg the mating habits and resultant aggression of birds.

- Clarify your low energy times each day (eg 3 to 5 pm, 2 to 4 am). Plan ahead and avoid doing important tasks in those times (if you can).

- Even just for one day, populate your thinking only with examples of positive actions around the world. The sites at 'Future Crunch', 'Fix The News' or 'Good News' can give you many examples. After reading the optimistic stories, predict one good thing likely to happen this year; and then, this century.

- Identify a personal habit (e.g., exercise routine, study schedule) and track its regularity over a few months. Predict future occurrences and deviations. Recognize and predict personal behaviour patterns to improve your time management.

CHAPTER 5 IN SUMMARY

- Natural cycles like bird migration, hibernation, and celestial events are highly predictable. These patterns have historically helped humans plan their lives and are still used today for forecasting various events.

- Human behaviour is largely predictable due to daily routines and biological rhythms, such as circadian cycles. Understanding these patterns can help predict daily activities and improve well-being.

- Animals often exhibit behaviours that indicate upcoming environmental changes, such as storms or even earthquakes. Their heightened sensitivity to natural cues makes them valuable indicators for predicting certain events.

- Human-induced climate change is a major predictive challenge, with potentially severe consequences. While solutions like renewable energy and circular economies are proposed, the transition is complex and fraught with difficulties.

- Advances in meteorology, seismology, and astronomy have improved our ability to predict weather, natural disasters, and celestial events.

CHAPTER 6: LIFE'S A GAMBLE

There are two kinds of forecasters: those who don't know,
and those who don't know they don't know.

JOHN KENNETH GALBRAITH

The Thrill of The Chase

Paul the octopus attained media celebrity status during the 2010 World Cup. He correctly predicted the outcome for Germany in all seven of their matches, including their play-off for third against Uruguay. His keepers would place two boxes that brandished the flags of the two competing teams into his enclosure, and the prediction was determined by which box he ate from first. In all, he correctly predicted 12 out of 14 matches in his career, a success rate of over 85%.[1]

Don't expect returns of 85% whenever you bet your hard-earned money. It is more likely to be the amount that you lose. After researching various gambling and predictive betting systems around the planet, I have one piece of advice for you: Please don't gamble. However, if you enjoy the thrill of betting, and you can afford to do it, and you are prepared to engage in research, and you know how to step back when you have reached your limit, then enjoy yourself. Sadly, too many gambling addicts do not know when to stop. The expression 'Hook A Whale' is commonplace within betting companies, and it refers to the gambler who often loses.

There are generally three types of gamblers. The professionals (perhaps less than 1% of all gamblers) approach the system from a strictly financial perspective, and research exhaustively, invest wisely, and treat it as serious work.[2] However, even those people have poor years when they lose much more than they make. Social gamblers (around 90%) set out to have fun, and generally make small profits or losses. The entertainment and socializing are as important to them as anything else. Problem gamblers (around 9%) compromise their lives in the desperate chase for the big win. When they spend grocery or rent money on a bet, they have a problem.

There are significant costs to the community from gambling addiction. Gambling losses in different countries are led by the US, with $116 billion each year. Australia is 5th on the list at $18.3 billion yet is only the 55th largest country in terms of population. At $958 each year, Australia has the highest losses per resident adult in the world.[3] This is averaged from all adults, not just people who gamble. At the individual level, the costs to you may include failed relationships, loss of employment, and even bankruptcy. The losses for society are just as problematic. In 2017, total gross revenue from the gambling industry in Victoria amounted to $5.8 billion. The tax revenue was $1.6 billion. However, the social costs came to approximately $7 billion.[4]

Gambling is about much more than simply winning some money. Reasons for betting might include the outright thrill of seeing a winning result, or the socialization with fellow gamblers, or the vicarious thrill of attending the event live. What plays a part more than anything else is the reaction of your brain to the gambling experience. Gambling distorts your reality, and can have as strong an effect on your reactions as abusive drugs.

Doing something enjoyable such as watching your team win, or by eating chocolate, releases a brain chemical called dopamine. Some researchers have conjectured that gamblers become desensitised to dopamine exposure, which causes them to seek greater risks to achieve the earlier highs. Further studies have indicated that gamblers do not necessarily experience their thrill from the specific win. The addictiveness comes from the uncertainty surrounding the entire experience. Pathological gamblers release greater amounts of dopamine even when they are losing, as compared to non-gamblers.[5]

Against The Odds

Casinos openly welcome people who are stimulated by chance events. If you believe in astute gambling, let alone intelligent investing, you will not walk through the doors of a casino. They are a profit-making business, not a charity. A 'house edge' is built into their machines, and the odds are that you will walk away with a little less money in your pocket. Slot machines have a 'house edge' between 1% and 15%. On a roulette wheel, that 'edge' is just over 5%. They will make about $5000 profit from every $100,000 that is bet.[6]

Prediction with the roulette wheel is near-impossible, even though James Bond always seemed to score huge wins. Too many bettors are swayed by the gambler's fallacy, in which they believe that past results affect future consequences. Just because the roulette wheel has scored 5 Reds in a row does not mean that the next roll is more likely to be a Black.

One math savant claimed that tiny imperfections in the roulette wheel could increase his ability to predict the correct number. If the wheel develops a slight imbalance, it can create a form of 'drop zone' in which the ball is more likely to fall.[7] However, if the table is perfectly balanced (and the vast majority are), then a win is pure chance, and only outright luck will lead to a profit for you. Because of the accumulation of the 'house edge', you will lose more the longer you play.

Casinos often lack clocks or windows, which makes it harder to stay aware of time passing. Some also provide free drinks, which means that your judgment will be impaired.[6] In a casino, there is no such thing as a lucky number that will get better results. When you are drunk, you tend to think otherwise.

You do not have to be drunk to be fascinated by numbers. The connections between numerology and luck have resonated with people all through history.[8] It is based upon the mystical connection between a number and some coinciding events. 3 and 7 are considered as lucky numbers in most cultures, although Chinese people prefer not to use 7. Thirteen is unlucky in most cultures, and Friday the 13th reputedly indicates bad luck. In China and Japan, 8 is considered lucky because their pronunciation of the word sounds similar to *prosper*.[9,10]

Four is another unlucky number in China, and citizens avoid having it in their address if they can. However, 4 is a good luck symbol in most of the Western world, because a four-leaf clover is considered to be fortuitous. There is no scientific validity to any of these beliefs, but if you are determined enough, you will believe that your lucky numbers might lead to an amazing fortune.

Lotto winners somewhere have probably claimed that they were guided to the winning numbers through divine intervention, a vivid dream, or some numerological system. A Lotto win is due to chance as much as anything else.

Most people will have a story ready to tell if they win. Those winners are then ready to immediately share the narrative about their system for choosing numbers, and so the mystique about number selection grows with each telling. Some Lotto players choose numbers based upon the previous luckiest numbers that were drawn, although the numbers featuring in each new draw will have the same chance of coming up as any other numbers.

It is possible to improve your chances of selecting six Lotto winning numbers. Buying a Systems entry with nine numbers instead of six on each entry lowers your chances of a Division One win to 1 in 96,965, which is over 80 times a better chance of winning than choosing six numbers only. A System 20 entry gives you the very worthwhile odds of 1 in 210, although it will cost you $32,064.[11]

20% of all Lotto wins around the world are achieved by syndicates. These help you to invest in the higher Systems entries. Just buying lots of tickets obviously improves your odds. There might be some merit in choosing numbers above 31, given that many people choose their family's birth dates. If your ticket then wins, it will possibly be shared with fewer people.

On rare occasions, some people have worked out mathematical loopholes with Lotto systems. In the early 2000s, a well-known couple in Michigan eventually won over $38 million with their mathematical system. If no one won, the game called Windfall would roll down each week until it capped out at $5 million. If that amount was not won, it was then allocated to a few lower winners who had most of the winning numbers. The husband, Jerry, calculated that if you purchased enough tickets, you were a guaranteed winner. Their final play cost them $712,000 which purchased 366,000 tickets. It took them four days at 12 hours every day to feed the tickets into a machine. The loophole has since been closed.[12]

The Power of Seven

The number seven is historically associated with many natural experiences. Seven is statistically the most likely combined number to be thrown with a pair of dice. There are seven colours in a rainbow; there were seven wonders of the ancient world; there are seven notes in the common musical scale; we have seven chakras; there are seven deadly sins; we have seven days in the week; there are seven seas; and we even have seven continents. Each phase of the moon – new, first quarter, full moon, third quarter – lasts for approximately seven days.[13]

Magicians know that 7 is more likely to be chosen when people are asked for a number between 1 and 10. Seven has significance in most major religions. God created the world in six days, and rested on the 7th. The Book of Revelation features 7 extensively, with 7 churches, 7 angels, 7 seals, 7 trumpets, and 7 stars. Seven heavens are outlined in the Koran. There are 7 higher worlds and 7 underworlds in Hinduism.[14]

Claudius Ptolemy of Alexander defined 'The Seven Stages of Man', namely: The Moon (0 to age 4), Mercury (ages 4 to 14), Venus (age 14 to 22), The Sun (age 22 to 41), Mars (age 41 to 55), Jupiter (age 55 to 67), and Saturn (age 67 on). Other writings since that time align with these seven stages.[15] Some people welcome the next stage, especially Venus and the Sun, while the more chronologically gifted are dragged kicking and screaming into the inevitable older stages. While the transitions might feel disruptive, they actually make your life more predictable. In each of these stages, your character is partly determined by the experiences you have had in your life. In rapidly changing modern times, those experiences can vary eg growing up with social media.

The overwhelming incidence of seven is important for someone who is calculating or predicting events in the future. A greater proportion of people are more likely to include seven in their calculations (eg with a Lotto win or a horse race), which means that the winning amount will be shared by a greater number of people.

Betting On Sports

Sports betting appeals more to many punters, partly because they simply enjoy the sport. The advantage of betting on most sports is that there are only two competitors or teams. Even then, if you don't win at least 55% of the time in a two-team sport, you will lose. So how can you predict the team that is more likely to win? You research exhaustively. What happened on previous games between these two teams? What's the winning record of each team at this specific venue? Which team was caught up in controversy that would affect team harmony during the past week? What are the experts predicting?

Machine learning software is available which allows you to enter variables such as the weather, previous results against the opposition, and many other indicators. Search for 'machine learning software for sports prediction', and you will find plenty of companies happy to help you out. Their systems are based upon building a classification model that rely on a training data set. As you feed data into the system, it can determine patterns and develop a probability of the result.

If you want to develop your own, then decide on the data set you want to use and draw it up on a spreadsheet. Give a value to each item in the data set. For example, if prior results between the two teams is a major indicator in your belief, then it might carry a weighting of 30% to the sum total of the data set. Between the two teams, you would allocate a total of thirty points for prior results. The remaining seventy points would be allocated to the other data points e.g. the weather, or the stadium in which the game is being played. Back-testing your own system from results in the past few seasons will help you to refine your model.[16]

After developing your spreadsheet, you then determine the probability of a team winning. You might decide that it is a 70% chance. If you can get odds of $1.90 or even better (which means you will receive $1.90 for every $1.00 you bet), then it might be worth the gamble. If the win will only pay you $1.20, then it is probably not the best use of your money. If you calculate that the favourite is well-priced even after you do your research, then it may be your bet of the day.

The same concept can apply to horse-racing, although there are many different systems that add to the complexity, including win bets, place bets, and trifectas. Beware the bet that you take because you just hope the team, or the player, or the horse, will win. That is pure sentiment, not intelligent betting.

While many complicated betting options exist in team sports (e.g., the first player to score), most bets focus on win, lose, or draw. If you are professionally serious about your gambling, then a predictive data set can provide you with some indication of the result. Your final decision still needs to be grounded in your own well-researched expertise. Remember that you are not the only person interested in a sports result. The owners of the team, as well as the coaching staff, the players, the bookmakers, journalists, and researchers all want to gain an advantage by predicting the result.

Thinking About the Stock Market

While millions suffered from COVID-19 in 2020, the Dow Jones Industrial Average also caught a serious infection. Between 12th February and 19th March, it lost 9362.90 points, which was a 31.7% drop. You can only imagine the amount of adrenaline flowing through investors during that time. A small number of economists had warned that it was very possible. However, most reputable experts are cautious about predicting where the market will be next week, let alone next year, given the endless factors involved.

Nate Silver talks about 'the signal and the noise' in his book of the same name. When you engage in predictive analysis, signals refer to data that clearly indicates the financial health of a company or the market in general, while the noise relates to myriad factors such as online rumours, or the endless small price corrections.

Nowhere is there more noise than with the stock market, given that it is the key arena for investors around the world. Share prices will always flow between boom and bust over a full cycle. For someone to predict that a recession is imminent does not usually take foresight. It is a given that it will happen somewhere in the next few years. No-one can pick these phases accurately. At best, experts may glean a general trend, although they are too often incorrect. John Kenneth Galbraith once said that "The only function of economic forecasting is to make astrology look respectable."

While market prediction might simply be wishful thinking for many, it has not stopped some people from developing theories on stock market movements. Back in the 1930s, accountant Ralph Nelson Elliott analysed over 70 years of stock data, and surmised that the market moves in the same repetitive patterns. Now referred to as the Elliott Wave theory, these patterns are supposedly affected by the emotional states of investors, which in turn are caused by outside influences such as media reporting. Does the theory work? Because of its subjective nature, most analysts do not agree with each other on their interpretations of the waves, although there are some who strongly argue its validity.[17]

Theories abound on the timescale for a full market cycle from one peak to the next. Those theories can stretch from 7 to 18 years. However, within that full cycle, you can be confident that there will be four phases. It will rise, peak, dip and bottom out.[17] The accumulation phase is at the bottom of the market, which is when most people wish they had never invested. Yet this is the best time generally to buy into the market. Unfortunately, most people have low confidence in shares at this stage, and will too often sell out.

The second is the Mark-Up phase, when the market becomes more stable, and prices are steadily rising. Astute buyers have moved by now, although the laggards will wait until the next phase, known as Distribution, before they buy back in at the peak. Sadly, for them, the eventual Mark-Down phase is when prices begin to drop again. Even worse, too many will sell in this stage, even when it means their shareholdings will lose in value. The key take-away with the stock market? Invest wisely for the long term, and stick with your investment process. The phases are going to happen anyway.

Investing Without Panic

The best approach you can probably take to investing is to learn how to control your brain's thinking when panic is imminent. You cannot control something like the stock market, but you might be a chance with controlling your own feverish responses to it. The secret is to not get caught up in your emotional state, but to become aware of what your brain is doing. Panic is usually associated with a surge of adrenaline, and it is not easy to regulate your thinking when you are in this mindset. Back in 1936, John Maynard Keynes wrote of 'animal spirits', and described them as a "spontaneous urge to action rather than inaction".[18]

Whether the market is in a raging bull mode, or a bearish negative sentiment, you can easily be caught up in these animal spirits, and feel that you need to act. The age-old advice of 'breathe deeply' is always worth the time it takes to do so, given that an oxygen-starved brain does not think as clearly. I can attest to being caught up in the mad rush for the financial exit. As a result, I have now learned to walk away for ten minutes (or sometimes ten hours), and do some worthwhile thinking before I press any buttons to sell or buy. The 'sunk cost fallacy' can compromise your judgement as well. Once you have invested in a specific share, you too often persevere with it because of the money already allocated.

Diversifying your investments, which involves you buying shares across a wide range of markets, has strong merit. It can smooth out your gains and losses.

As well, jumping in and out constantly will frazzle your nerves, and it may even not lead to an improved profit. Just remember that, with rare exception, shares are not a quick fix investment. Warren Buffett probably said it best when he opined that "The stock market is a device for transferring money from the impatient to the patient."

If you are after something or someone to support you investing in the market, you have two basic options. Find a registered financial planner who has demonstrated a long-term proficiency with money management. Otherwise, do your own intensive learning, and begin to trade with an online market simulator for at least six months. It will cost nothing, and yet will provide you with some powerful lessons. Predictive analysis systems flood the market, although you need to cross-reference online about their reputation.

When your brain craves a little more excitement, you could always dabble in the cryptocurrency market. The older and wiser investors such as Warren Buffett refuse to touch it, and correctly point out that it has no real market value. Other more adventurous investors delight in supporting this alternative payment system to global banking. Bitcoin's initial value was less than $0.001, and the first purchase was for two pizzas delivered from a Florida restaurant for 10,000 BTC. In October 2021, it reached a high of US$66,975 before rapidly plunging in value to $18,000 in June 2022. At the time of writing (July 2024), it was around US$61,117.[19]

If you can accurately predict Bitcoin progress well into the future, then you deserve any success you might have. You must enjoy roller-coaster rides, and especially when you are willing to take your chances with all the possible scams as well. In the cryptocurrency world, there are numerous frauds, partly because most people know little about it.[20] This makes the novice investor an easier target.

Beware The Futures Scams

The global scamming industry is lucrative, given that over $485 US billion was lost to global scammers in 2023 alone.[21] Note that this is only covering the reported scams. Over 30% of victims never tell anybody. The types of scams can extend from information request or phishing scams to financial advice scams, and to buying and selling scams. The sad reality is that greed blinds people, and brings out the worst in their 'stupidity' gene. Always remember: If it is too good to be true, it invariably is. Let's explore how you might be tricked by scammers who specifically claim that they can predict future results.

The most common are scammers who declare they have earlier predicted a recent stock market downturn. When you read the email or the report that they wrote years previously, you can see that they did predict what occurred. The online report is clearly date stamped. What they neglect to tell you is that they wrote a whole variety of reports at the same time which cover the full gamut from a significant downturn to a sizeable upturn. The unwanted predictions are deleted online, and simply are not mentioned in their promotions. They then claim to have a lucrative system that will help you with your predictions. There are many variations on this multiple reporting.

Many scammers use the 'stopped clock' syndrome. Even a stopped clock is correct twice a day. The scammers state month after month that the market will nosedive. They will eventually be correct, even if it takes years to unfold. This reaches further levels of disbelief when the mainstream media delight in reporting this supposed prediction. The same media rarely point out all of the other predictions that were incorrect.

The negativity is heightened by the emotional appeal common to these forecasts about the impending doom. If their predictions feature in a video, the background music is as fearful as any horror movie you have ever watched. Rest assured that whoever is doing this negative forecasting will benefit from the fear that it creates.

There are plenty of variations on this predicting format. One is that the scammers record ten or more videos about an upcoming event such as the US Presidential elections, stretching from one extreme result to the other. They then place them on to YouTube, which records the date that the video was taken. The scammers do not actually release the videos for public viewing. After the event, they launch the one video that was closest to the result that eventually occurred. Cryptocurrency scammers work with the same process. The final single video then becomes part of the marketing for their prediction betting system.

Horse-racing prediction systems appeal to the struggling punter who would love to consistently have some wins. With one system, the scammer acquires a mailing list of perhaps 2000 punters who have previously purchased a horse racing betting system. The scammer chooses a race next Saturday that features ten runners and sends the name of each horse to 200 different punters on the list. At the end of that Saturday, a specific 200 people are impressed that the scammer has chosen the winner.

Another race with ten horses is chosen the following Saturday. On this occasion, the name of each horse is sent to groups of 20 people in the 200 remaining. At the end of that second Saturday, 20 people out of the original 2000 are highly impressed with this new system that has predicted two winners in a row. Those 20 punters then receive persuasive marketing material encouraging them to purchase this amazing but expensive system.

Melbourne racing identity Bill Vlahos defrauded over $17.5 million from punters between 2007 and 2013. In a club that he named 'The Edge' (which was a Ponzi scheme that simply defies belief), 1800 'investors' trusted Vlahos to invest their money in horse races. One bettor alone transferred $1.2 million into The Edge bank account. Friendships were broken forever, given that most of the investors were sourced by word-of-mouth.[22]

Many similar schemes around the world have involved scammers who simply announced on social media that they were a betting company, and that they would be offering amazing odds on a specific horse that was racing on the next Saturday. While the market price may have been $2.20 for a win, the bogus bookkeeper entices gamblers with odds of $6.00 or more. If the horse actually wins, the scammer disappears from view.

Scams that are supported by future AI will test our collective resolve much further. Imagine receiving a phone call from a close friend, who explains that he desperately needs a loan. His voice sounds authentic, and he is engaging in conversation with you. Perhaps one of your children is travelling overseas and sends a video requesting some money to be sent to a new account. Both instances can already be artificially developed by scammers who have created avatars to mislead you. Trust your gut feeling, and cross-check the authenticity of the original call.

Predictive Exercises to Expand Your Predictive Ability

- If you have to bet, calculate the potential win or loss and compare it to the probability of each outcome. Determine if your betting decisions are based on sound risk assessment or on emotional impulses.

- Develop % probabilities (rather than yes/no predictions) for sporting event results. Work out the best factors to consider when developing those probabilities.

- Maintain a journal of your emotional state before, during, and after gambling sessions. Record your feelings, stress levels, and any physical symptoms (e.g., increased heart rate). Identify patterns in emotional responses and understand how gambling affects your mood and mental health.

- Choose a sporting team you do not follow. List the 5 most reliable indicators (eg weather; previous results against the same opposition) for predicting a win in their next game. Then apply your indicators and calculate their chances in that game.

- Analyse your level of panic when the stock market drops. How could you improve your response in order to benefit from the fall?

- Research the long-term trends of a specific type of shares eg technologies. Read the data exhaustively. Then predict the movements of those shares for the next year. Follow the prices carefully, and monitor how closely your prediction was correct.

- Research three recent scams, and tell your family and friends about them.

- Study concepts like loss aversion, the house edge, and sunk cost fallacy. Reflect on how these principles have influenced your past gambling behaviour. Apply this understanding to avoid common psychological traps in future gambling decisions.

CHAPTER 6 IN SUMMARY

- Gambling is often more about the thrill than financial gain. While some professionals approach it methodically, most gamblers do it for fun or fall into problematic patterns, leading to significant personal and societal costs.

- Gamblers frequently believe they can predict outcomes or find patterns, such as in roulette or numerology. However, these beliefs are typically unfounded, and casinos are designed to ensure the house always has an edge.

- Predicting sport events involves extensive research and data analysis. Machine learning and statistical models can help, but success requires careful consideration of various factors and disciplined decision-making.

- Predicting stock market movements is highly complex and often unreliable due to numerous influencing factors. Understanding market cycles and maintaining emotional control are crucial for long-term investment success.

- The gambling and investment worlds are rife with scams. Scammers use various tactics to deceive individuals, including fake predictive systems and AI-generated communications. Staying informed and sceptical is essential to avoid falling victim.

CHAPTER 7: INTUITIVE PREDICTION

It is through science that we prove,
but through intuition that we discover.

HENRI POINCARE

Are Psychics for Real?

There are nearly 94,000 psychics businesses in the US.[1] That does not account for the many other people who profess to have psychic ability. No evidence-based data exists on the percentage of legitimate psychics in the population, assuming that some can indeed demonstrate other-worldly abilities. According to the Pew Research Center, about 4 in 10 U.S. adults believe in psychics, so there is certainly a market for these claimed abilities. Those talents may include fortune-telling about your future financial or romantic life, and the professed ability to provide support for your grief or loneliness. This begs the obvious question: When it comes to predicting the future, do authentic psychics really exist?

I have met someone I believe to be a legitimate psychic. This admission is coming from someone who needs scientific validation for everything. I also acknowledge that one experience does not bestow scientific credence. Years ago in a small Australian country town, I was introduced to a lovely elder gentleman (let's call him Karl) during an evening dinner. Several local people had pre-warned me to expect the unexpected, and so I was curious about the upcoming connection.

When I first met him, I was fascinated by his highly unusual ice-blue eyes. They gave the distinct impression that he was looking right into me. Karl was of European ancestry, and he later mentioned that he had been signed up under four official secret service acts during World War 2. He reputedly knew when the bombers were coming, and where to find hidden bodies.

Over the course of the evening, he referred to matters in my personal life that were impossible for him to have researched. No one else, without exception, knew this information. He also predicted several major experiences that he claimed would occur in my future life, which eventually came true. I felt steady and calm about this unveiling. As the time neared midnight, his wife said to him: Karl, give Tony some energy. He moved around behind my chair and placed his hands gently on my shoulders. A few seconds later, my entire body was filled with the most glorious surge of energy.

The next day, I began my car journey back to home. A few minutes after the highway drive started, I came to a long upward incline. Suddenly, two huge trucks sped side-by-side towards me over the rise of the hill, with one obviously intent on passing the other. They were less than two seconds from me, and they were taking up the entire road. The space beside the road was too narrow to fit a car, and there was a sharp slope off the edge.

What was my first thought? Interestingly, it was that Karl had predicted some events in my future, and that this therefore could not possibly be the time for me to die. I calmly turned my car sideways down the steep decline beside the road. The blast of wind from the two trucks rushing nearby rocked my car. A little further ahead, I was able to carefully drive back on to the highway and proceed on my journey.

What is your own experience in these matters? If I had interviewed you, would you have been able to provide a powerful anecdote? In the rest of this book, I have sought scientific verification on all predictive processes. There is none for psychic ability. Does that mean that it is a fallacy? Not necessarily. Some things simply cannot yet be explained, given that the present rules for classical mechanics break down at the atomic level. Perhaps the revised understandings on quantum physics by 2150 AD will shed further light on this possibility.

Quantum Psychics

One conjecture is that legitimate psychics exist in a different time cycle. While the rest of us live in a linear world, they might move through a quantum existence, and engage in a parallel universe that is in a slightly altered time. This 'time' is perhaps not the linear constant that many of us are led to believe. Einstein once wrote that: "People like us, who believe in physics, know that the distinction between past, present and future is only a stubbornly persistent illusion."

Maybe a true psychic can manage to observe events in each of the three time zones, or in some form of parallel universe, or can operate without the mental construct of linear time.[2] The emphasis is on 'maybe'. While there may be vague indications of a multiverse, it is not substantiated by present day physics. It is more philosophical than physical.

Theories on physics are a work in progress. Just when we think that we have discovered most that there is to know, we then are confronted with explaining a concept such as quantum mechanics.[3] Essentially, it is a study dealing with the interaction between sub-atomic particles. This study has significant implications for the world up ahead. One may be quantum computers, which would massively boost technology capacity.[4] Another might be at the biological level, which may help to explain scientifically how to engage in psychic activity.

I still do not believe that most self-claimed psychics can really foresee the future. I suspect that their talent rests in sensing someone's emotional response to what is said to them. For now, I will accept that a select few really can predict some aspects of the world up ahead. If they can, their talents may be astonishing. Psychics reputedly possess such abilities as prophecy, energy medicine (healing with your own mental or spiritual energy) and even levitation. ESP (Extra Sensory Perception) is a little more grounded and includes such gifts as clairvoyance (seeing things anywhere else in the present), precognition (having some knowledge about future events) and mental telepathy.

The adherents who believe in psychic ability sometimes claim that any one of us can see the future, but that we have been convinced to downplay that gift by a disbelieving society. And if scientific analysis takes place, believers maintain that the negative energy in the controlled testing environment weakens the ability of any psychic. For them, such testing is not valid. Regardless of the lack of proof about clairvoyancy, some police services around the world have reputedly sought out psychic evidence available about a crime.[5] The dilemma for the police is that they eventually need to present legal and irrefutable evidence.

Following Your Intuition

If you struggle to comprehend psychic ability, then perhaps you are more open to intuition. The very word can be broken down into Inner and Tuition, which means that you are learning from within. With your intuition, you make mental connections with previous knowledge that is stored in your amazing brain, and then determine the probability of future events occurring.

Intuition is an indication of higher intelligence. You assess all obvious cues, cross-reference with your memory, and then develop a reasoned conjecture about the future possibilities. When you hire a new person for your business, or you embark on a project in the first place, or you meet a prospective partner for the first time, you are probably using your intuitive ability.

A more basic version of intuition is called gut instinct, which is hard-wired into your body and brain. It refers more to coping with difficult situations, and with how your brain responds in order to survive the event. We sometimes refer to this experience as extrasensory perception or a 6th sense. With gut instinct, you experience such effects as the fight or flight syndrome, in which your brain and body involuntarily respond to the situation. In ages past, this was to run from the sabre-toothed tiger. In today's world, it might be to not meet up again with someone you have just met by chance. You just know, although listening to your gut needs some degree of confidence in the process.

Psychologists propose a dual-process theory with decision-making. We connect with both our intuitive (tacit or experiential) and our analytic (rational or deliberate) abilities. Malcolm Gladwell describes the two processes as blinking (the intuition) and thinking (the analysis). The blinking is done when a decision needs to be made immediately, and the thinking occurs when there is time to ponder the options. They are both valuable concepts, although Gladwell contends that *"There can be as much value in the blink of an eye as in months of rational analysis."* [6]

Intuition can give your brain a rest. Rather than having to think extensively about something, you simply jump in with your intuitive response. Is it always valid? No, because this quick response can be tainted by subjective judgement that is affected by your biases and lack of contextual knowledge. However, modern life and work in particular can leave you with little time to ruminate at length on an issue.

Intuitive responses may save you both time and energy, and especially when you have sharpened your intuitive pattern-making ability. When you are an expert at something, it is partly because you have experienced many situations that collectively built up in your memory. If you then face another similar pattern forming, you can intuit what is likely to happen.

In *Thinking Fast and Slow*, Daniel Kahneman outlines three principles that need to be satisfied before you can trust your intuition.

1. There are boundaries on the environment that you are making an intuitive prediction on (it is a slow and unchanging system)

2. You have a lot of practice in making these intuitive predictions (there is a large sample size)

3. You receive immediate and accurate feedback from your prediction[7]

Perhaps you have just begun a new relationship, and you are already predicting the wedding in your own mind. That is not intuition. It is hope. None of the three conditions apply in this case (unless you have had lots of practice with new relationships).

How can you enhance your intuitive ability for now? Find your natural pace. Sense what is all around you, rather than rushing everywhere without observing what is nearby. Meditation would be a great start and can help you to rediscover that natural pace. Acknowledge the messages that your body is giving to you. Learn to sense those messages rather than ignore them. Be aware of what a constant technology overload may be doing to you. Learn how to take some time away from electronically artificial stimulus.

Go back to physically handwriting in a journal. Too many people have forgotten what it is like to hold onto a pen. This utilizes your fine motor skills and stimulates your thinking.[8] Look to nature for indicators of intuition. Animals do not consciously think when determining what to do. Observe your pet dog / cat / fish / parrot as they respond to stimulus. Animals act intuitively, and often in synchrony with their flock or group.

Everything Is Connected

Think of the last time that you viewed a flock of birds high in the sky, changing direction is though they were a single entity. Large schools of fish likewise swim as if they were one. When people join a choir or orchestra, they often experience this sense of connection. Heart rates sometimes synchronise, and body movements mirror the others in the group. If you find the concept interesting, then you might be intrigued by the theory of 'morphic resonance' proposed by UK biologist Rupert Sheldrake.

According to Sheldrake, a morphic field is a form of energy that permeates the biology of any living things. All monkeys, all crows, all humans, and any other creature you wish to name, innately understand what all other members of its species have previously learned throughout history. Sheldrake attests that morphic resonance is "the idea of mysterious telepathy-type interconnections between organisms and of collective memories within species."[9] This is meant to explain that doing newspaper crosswords is easier in the afternoon (given that lots of people have already done it by then), and that all monkeys around the world eventually intuit how to wash a dirty potato in a stream of water once one has initially washed a potato.

There is a plethora of circumstantial evidence for Sheldrake's theory. One interesting experiment with rat learning began in the 1920s at Harvard and continued for several decades. We need to partly sympathise with rats, given how often they are used in scientific experiments. In this case, they learned how to escape from a water maze. Later generations became faster and faster. Once the Harvard rats could escape up to ten times faster, other future experiments in Edinburgh and Melbourne found that their rats basically started from the point that the Harvard rats finished off. The Melbourne rats continued to improve further.[10]

Suffice to say, the science establishment is not impressed, and numerous articles can be found that refute morphic resonance, referring to it as a pseudoscience at best. Scepticism is usually the default position with Sheldrake's work. The clinical evidence for his theory is mixed. This once again is a concept that is not easily proven, and yet we might sometimes intuit it to be partly valid.

The possibilities, if this theory is ever accepted, will alter how we think of humanity. Imagine if a collective consciousness between all the world's population was ever substantiated and practised. Predicting future courses of action might be determined by some form of global brain. Perhaps our present electronic communication devices are a precursor to how we can eventually connect via some form of transhumanist mental telepathy.

Empaths In the Real World

Empaths[11] would be among the first people to engage in any version of mental telepathy. An Empath is a person who can deeply sense emotions in others. There are overlaps here with people who are intuitive, although there is a distinction between them. Empaths are intuitive, although intuitive people who might foresee the future are not necessarily deeply empathic. There is also a variance between an empath and a person who is introverted. The latter are just reserved, and may or may not be strongly empathic.

Empaths deeply feel the emotions of others, and exhibit much more empathy than average, which can become both a blessing and a curse. The blessing is their ability to profoundly sense the emotional state of a nearby person (or even one on the other side of the planet). The curse can be the burden of sensing the issues of other people.

Empaths are often excellent storytellers. They are invariably good listeners and are curious and highly astute. They need constant stimulation, and they support the underdog. Unfortunately, they are not always capable of setting boundaries, and they can feel overwhelmed at interacting with too many people. To cope with their high levels of empathy, they may need to spend more time in nature. They may also need to learn how to say no when they are asked to help someone, given the emotional pressure that it might place upon them.

In *The Empath's Survival Guide*, Dr. Judith Orloff describes Empaths as "emotional sponges." Orloff claims that there are four different types of em- paths.[12] The first is the Dream Empath, who often have vivid dreams that they can remember the next morning. They can even instruct their brain to dream of specific issues that need to be resolved.

The second is a Telepathic Empath, and they can read feelings and thoughts in others in the present time. This Empath can easily become overwhelmed by the collective thoughts of everyone around them, including people they do not know. The third is the Earth Empath who is intimately connected to the natural world. These people strongly feel any changes in the weather and may be more affected than others by SAD (Seasonal Affective Disorder).

The fourth group, according to Orloff, are Precognitive Empaths. These people reputedly experience premonitions about the world up ahead, whether when awake or in their dreams. There are some concerns about having a gift such as this. For example, what would you do if you sensed that a friend was going to have some difficulty up ahead?

It is advised that you think carefully about whether your friend would be best served by hearing what you can offer. Perhaps you can check with your own intuitive judgement on what to do. The concern is that you might become motivated by ego, and just want to boast about your ability. This does not pay respect to your precognitive gift, or to your friend's need.

Scientists have established that we have 'mirror neurons' in our brain, and these may help us to 'mirror' someone else's emotional state.[11,13] One conjecture is that empaths might have a greater number of these neurons, which may explain their ability to deeply sense what others are thinking and feeling. Some unusual effects may be experienced with this ability. One is that the empath may feel some form of sensation on her own skin when she watches someone else being hugged or touched.

Time Travel Unraveled

Intuition and empathy might allow you to sense some of what is up ahead, although the simplest way of predicting the future would be to invent a time machine. Just imagine. You take a quick trip one week ahead of today to watch the US Powerball draw, write down the winning numbers, and head back to the present time to submit your entry.

There are just a few issues with this process. One is that you cannot be in two places at once. Your own self up ahead will not be happy to have a younger self in the same time period. Another is that you may be altering the events that would have taken place up ahead. The most obvious issue is that we are a very long way - if ever - from constructing a machine that travels across chronological boundaries. Time travel into the future, at least for the non-psychics amongst us, appears for now to be unlikely.

Perhaps it might be easier to travel back in time. You could redress some of your worst past mistakes. Will this ever be possible? Probably not, although we are not sure what people in 500 years might be able to do.[14] Marty Mc-Fly may well have performed the effort with his DeLorean car in *Back to the Future*, but that remains in the fantasy science fiction category. The *Star Trek* series featured time travel by making use of parallel universes. Some scientists conjecture on future possibilities that involve astonishing speed, or travelling through wormholes. You will probably not live long enough to see these far-fetched possibilities become a reality.

Another option for space and time travel might involve black holes, which create gravity so strong that light cannot escape. As you neared the black hole, time would become slower. As you moved away, the time would pass more rapidly.[14] Some future advanced technology might allow you to slip into some form of time warp near these black holes. You would need an awe-inspiring spacecraft to cope with the gravitational forces.

Brian Cox, an eminent UK physicist, once claimed in his usual tongue-in-cheek manner that forward time travel is vaguely possible. His explanation focused on the proven fact that moving clocks run slower than stationary ones. If you went for a run wearing one watch, and left another behind, the moving watch in your hand would slow down by the time you returned. However, the difference is so negligible that even an atomic clock would hardly measure the nano-second variance.[15]

University of Connecticut physics professor Ronald Mallett once proposed an intriguing device that could receive messages from people in the future. Given that light alters time, he considered using a rotating stream of light that could twist space and time. The message would be composed of neutron spins that would transmit binary code from the future to the past. The messages could only go one way, so you would not be able to return to sender.

The cost of this device? $250,000. It sounds like a bargain, given the amazing things we might be able to learn. Most other physicists have demonstrated both skepticism and mirth about the option.[16]

Predictive Exercises to Expand Your Predictive Ability

• For a week, record daily decisions you made that were based on gut feelings. At the end of the week, review the outcomes. Were the instincts accurate? How did they compare to decisions made through deliberate analysis?

• Visit a 'psychic' and document the predictions given. Track these predictions over a year to see how many, if any, come true. Analyse the validity of these psychic predictions in a structured, evidence-based manner.

• If you lived in 2198 AD, and could send a message back to now, what would you say to today's global inhabitants?

• Use card games like poker to predict outcomes based on patterns and probabilities. Reflect on how intuition plays a role in your decision-making process.

• Be aware of when you most often use your intuition eg when employing a key person; or deciding on a critical course of action; or responding during a meeting. Think of how your intuition could benefit you (and perhaps others) over the next week.

• Scan a room (or a train, or a marketplace) full of people, and sense how specific people are feeling. Think to yourself: What awaits some of them in their near future?

• Keep a pet dog or cat. It relies on its own version of gut feeling. Learn how to tune in to its reactions eg when someone is approaching where you live. It can heighten your own senses.

• What would you love to ask your older self? Choose an age that would hopefully have the wisdom to respond with deep meaning. Which clarifying questions would best help you to traverse the path between now and then?

- Write a letter to your future self with predictions about your life in 5 or 10 years.

- Set a reminder to read the letter at the chosen future date and evaluate the accuracy of your predictions.

- Invent a time machine (!!). You can then travel into the future, look around, and return to tell the rest of us what to expect up ahead.

CHAPTER 7 IN SUMMARY

- The existence of legitimate psychics is debated, with anecdotal experiences and some belief in psychic abilities. However, there is little evidence-based validation for these abilities, leaving the question open to interpretation and future scientific exploration.

- Some suggest psychics might operate in a different time cycle or parallel universe, potentially explained by quantum mechanics. Theories like morphic resonance also propose interconnected learning across species, though these ideas remain unproven and controversial.

- Intuition involves making quick judgments based on subconscious processing of past experiences and patterns. While not always accurate, it can complement rational decision-making, especially when time is limited.

- Empaths possess heightened sensitivity to others' emotions, which can be both beneficial and overwhelming. Precognitive empaths reputedly experience day or night dreams that predict aspects of the future.

- While time travel remains a theoretical concept with numerous scientific and practical challenges, the idea continues to fascinate. Exploring concepts like black holes and advanced technologies may offer insights, but practical application is far from reality.

CHAPTER 8: EXPONENTIAL FUTURES

*If I asked people what they wanted,
they would have said faster horses.*

HENRY FORD

The Reality of Exponential Change

In 1965, former CEO of Intel, Gordon Moore, predicted that there would be a doubling each year of the number of components per integrated circuit.[1] This doubling effect, commonly known as Moore's Law, has become a cliché to represent accelerative change. He surmised that this rate of growth would continue for at least one decade. It certainly lasted much longer than that, although some specialists now contend that it has reached its technological limit. Others believe that with the continued miniaturisation of computer chips, Moore's Law may still be valid until the mid-2030s.

Exponential changes are now evident in many aspects of our lives. You may remember the excitement in 2001 with the sequencing of the human genome. It cost $100 million and required nine months to complete a sequence. Twenty years later, the most advanced generation sequencer performed the same process for around $100 and completed it in one hour. This is one million times cheaper, and 6480 times faster.[2]

If you thought that prediction was difficult up to now with our present pace of change, then consider the likely exponentialities of life in the 21st and 22nd centuries. Change may become so rapid that we will need to contemplate it in weeks and days, rather than years. For someone to accurately foresee what events might occur in 2150 AD is infinitely complex, given the number of transformational changes that will take place between now and then. We may witness more total developments just in the next few decades than we have seen in all of human history.

There is nothing new to this concern for rapid change that appears to make the impossible come to fruition. *Future Shock* author Alvin Toffler wrote about this acceleration back in 1970, yet it has continued unabated. Google's chief futurist Ray Kurtzweil refers to an inevitable 'singularity', in which exponential change takes place every split-second. It is beyond comprehension how our present-day brains will cope with this rapidity. In The *Coming Wave*, Mustafa Suleyman and Michael Bhaskar discussed the monumental scale of technological advance. They wrote that "We really are at a turning point in human history."

Thankfully, not everything is exponential, and your brain will be grateful for that. Turning up to your parent's house for lunch every Sunday occurs on a linear timeline. Each 7-day cycle in your life is as long as the next. You are ageing (depending upon the way that you look after yourself) at a relatively constant rate. With some aspects of our linear lives, prediction is made easier by what is consistent and steady. The dilemma for our 21st century society is that many other aspects of life are accelerating faster than we naturally age and think. To predict many things in such a tumultuous time is problematic, if not impossible.

British science fiction writer Arthur C. Clarke proposed Three Laws of Prediction[3] in the 1960s, and they all refer to the manner in which we view the future in terms of impossibility. The laws are:

1. When a distinguished but elderly scientist states that something is possible, he is almost certainly right. When he states that something is impossible, he is very probably wrong.

2. The only way to discover the limits of the possible is to venture a little way past them into the impossible.

3. Any sufficiently advanced technology is indistinguishable from magic.

The third law is well-known, but the first two also have special merit. What might you consider to be 'impossible' or 'magic' in the future? What about: Turning invisible; living forever; being physically able to teleport to any part of the earth in a few seconds; having all nations around the world prepared to work together? They all sound ridiculous, don't they? Although it is beyond our scientific or practical understanding for now, each of these conjectures may become reality over the next two hundred years.

The Rubberband Effect

Here is a time-honoured process for training your brain to cope with exponential change. With the Rubberband Effect, you stretch your thinking far beyond where it might normally venture. Once you have experienced this cognitive stretching, anything else you predict is going to feel more comfortable. For example, if you wanted to generate some innovative ideas for a possible side hustle business, you might begin by listing some highly improbable options.

These might include: Offering tourist tours of the Moon; helping your clients to become a World Leader for a day (take your pick of the country); providing food for the mass market that has been harvested within the past 24 hrs; replacing everyone's clothes with invisible cloaks. As you can imagine, the list of improbable options would be endless.

Why engage in a process such as this? It stretches your thinking to then address other more realistic options. Those could include: tourism tours for star-gazing; selling masks of world leaders for people to play-act at being important; developing a business that sells locally grown fruit and vegetables (perhaps to the nearest 100 houses); designing camouflaged clothes that can blend in with different backgrounds.

Let's combine the Rubberband Effect with Arthur C. Clarke's 2nd Law, and venture into the 'impossible'. Here are a series of transformational provocations for life in the mid and late 21st century. My intention is to sufficiently stretch your thinking about the future, which can help you to develop your own set of predictions more capably. Even just for five minutes, I often use this process myself before embarking on research and prediction with a specific issue.

PROVOCATION ONE. Hybridization Between Humans and AI

Supporters of the transhumanist movement[4] argue that AI will progress at infinitely faster rates than normal human progression, and that our only hope of coping with that increasing disparity will be to augment our own capabilities. Scientists conjecture on the singularity up ahead, with rapid change occurring every nano-second.[5] By the mid-2040s (although the predictions for this timeframe vary), change will be so rapid that we will struggle to maintain pace with these advances. The human body and brain may need to be dramatically enhanced with a combination of reverse brain engineering, artificial intelligence, and genetic engineering.

While at first this sounds preposterous to most people, remember that we already augment our bodies in many ways. The list includes heart pacemakers, plastic surgery, Cochlear hearing implants, bionic eyes, and body suits that support paraplegics to walk. Future options will include neuro-prosthetic brain chip implants to lower depression, brain–computer interfaces helping us to control household devices with thought alone, and internal nano-devices that eliminate cancer cells.[6]

As part of this quest to enhance humanity, some version of a BrainCap may eventually supplant mobile phones. Your thoughts will be transmitted (when you choose to do so) to another BrainCap user. This wearable device will also allow you to access the internet at that future time. All human history would be available to you with your BrainCap. Surgeons may eventually implant some version of this device inside your brain. The initial human 'guinea pig' would need to be very brave.

Brain sensors will likely be included in future smartwatches and headsets, providing early diagnosis for conditions such as depression and even dementia. Both Microsoft and Google are rumoured to have patented wearable brain devices that can decode thinking and access the internet. At that point, you would be much more capable of calculating future events, given your new-found capacity to research anything available online.

As this BrainCap tech is refined further, we may eventually realise that this thought transfer technology was created to prompt us with what we can innately do ourselves. Some people may one day discover that they can do everything that the device can do (send thoughts to others, or connect with anyone anywhere in the world), without wearing it. At that point, we would enter a remarkable new era in human relationships.

Issues: Transhumanism has many detractors, and for good reason. Determining the limits of body or brain enhancement is a very slippery slope. We still don't know enough about the amazing human brain to even begin to replace or enhance it in any form. Significant caution would be necessary with any brain augmentation. Your brain is likely to reject AI that is designed to replace its functionality. As well, the world already has endless concerns with its wealth divide, and not just between developed and developing countries. Even the developed nations will have to contend with the inequities between the rich and poor. Would only the wealthy be given access to this augmentation?

Probability rating of continued minor enhancement of the body/mind by 2075: 100%

Probability rating of extensive transhumanist application by 2075: 50%

Probability rating of some version of a BrainCap by 2075: 75%

Probability rating of transmitting thoughts without using the BrainCap technology: 5%

PROVOCATION TWO. Human Longevity

The numbers reveal a compelling human longevity story. In the mid 18th century, life expectancy at birth was between 28.5 and 32 years. The 1900 world average was 31-32 years, and the 1950 world average between 45[7] and 48 years. The 2019-2020 average ranged between 72.6 and 73.2 years.[7] Even though the earlier data was affected significantly by high infant mortality rates, our global longevity has more than doubled in just a few hundred years. The search for eternal youth was a constant theme in Roman and Greek mythology. This search will continue for as long as humans exist. Will we ever reach a stage in which immortality is a common occurrence?

A physical eternal life sounds far-fetched, although the 1st person to live past 150 years is probably already alive today. Around the world, life expectancy increased by six years just between 2000 and 2019.[8] Even if that rate of increase stays constant, we will be close to an average of 100 years of life as we enter the 22nd century.

There are many beneficial reasons for this longevity, including: reduced child mortality, healthier lifestyles, improved health care and better hygiene. Some dramatic medical advances may lead us closer to immortality. These might involve internal nanobots that constantly finetune our body functions, or vastly improving systems for boosting resistance to any disease.

Immortality sounds impossible to most of us at this time in history, although you can already live forever in artificial ways. In 2D video form, you can appear on multiple screens worldwide, now or in the future after you die. Three-dimensional virtual reality (VR) takes the use of stored images a step further. Concerts have featured a 3D VR representation of a deceased singer performing 'live' on stage.

MIT Media Lab and Ryerson University have announced a collaborative venture exploring what is termed augmented eternity.[9] Your Intelligent Assistant, or IA, might maintain your social media feed and online life after you physically die. Haunting from beyond the grave will take on a whole new dimension. Your IA would update your information on social media according to new online experiences and prevalent circumstances.

A further step towards immortality may involve your Intelligent Assistant being implanted in a humanoid entity after your physical body dies. In essence, an artificial copy of your brain would be placed into this robotic body. The humanoid would then continue to live in a way that reflected your personality, habits, interests, and even ethics.[9]

Issues: Equity again. Will only the wealthy afford to live forever? Another issue: Although world population would previously have levelled out around nine billion, the planet would soon be overwhelmed if we had fewer deaths than normal.

Probability rating of maintaining an eternal online presence by 2075: 100%

Probability rating of some people living past 150 years of age by 2075: 80%

Probability rating of some people physically living forever by 2075: 1%

PROVOCATION THREE. Water and Food Security

Would you fancy a meal of dried seaweed? In Japan, various algae and seaweed are popular foods and can be grown in both salt and fresh water. What about a selection of fried insects? By mid-century, your menu choices also may include false bananas, wild grains and cereals, and lab-grown meat, which smells and tastes the same as 'real' meat and is just as nutritionally valuable. We may need these alternatives. The FAO (Food and Agricultural

Organisation) believes that global food demand may increase 70% by the middle of the century.[10]

This will be driven by improved economic standards and rising incomes in areas such as Latin America, Africa, and Asia. This increase in demand contributes to other issues. Rotting food in landfills create the greenhouse gas methane which heats the atmosphere up to 80 times faster than carbon dioxide. Climate change will make growing and harvesting cycles less predictable.

Although we may end up eating more insects, their diversity and numbers on the planet are declining. Insects pollinate food crops and serve as food for many other animal species. At present, land degradation, fertilizers, and general climate change are driving insect populations lower.

The amount of food waste is monumental. Around one-third of all food grown on the planet is thrown away because of inefficient storage and packaging. To further complicate this, a fifth of edible fruit and vegetables, known as ugly food because of its deformities, is less desirable for most consumers.[11] It makes no difference once it is in your stomach. The world produces sufficient food for everyone, and yet over two billion people go hungry every day.

What can be done about all of this food wastage at the personal level? Develop more innovative methods of packaging food to maintain freshness and reduce damage. Buy only what you need. Take less food to family gatherings. You alone are not meant to feed everyone in attendance (unless this has been predetermined). Compost as much of your food waste as possible.

The disquieting opposite prospect is that the planet may experience food shortages in the future. If most of the northern hemisphere's chief growing regions experience severe weather conditions at the same time, harvesting loads may be severely curtailed. The world usually relies on a form of global smoothing. Poor crop production in one country can normally be balanced by decent crop yields in another. Failing to grow crops collectively in a single year would turn our surplus into a dramatic shortage. Weather conditions that are accentuated by global warming increase the likelihood of these patterns emerging.[12]

We have similar issues with water. The world has always had the same amount of water over billions of years, given the global water cycle. We have plenty of water in our oceans, rivers, lakes and in the atmosphere. The issue is that only 1% of that is drinkable. 2050 AD will be a landmark (or watermark) year. Water demand will increase up to 50% by that time.[13] The number of people living in river basins under severe water stress is projected to reach 3.9 billion, totalling over 40% of the world's population. Compared to the 2020s, five times as much land may be in "extreme drought" by 2050.[14]

What can be done about this? Desalination is one of the first responses to that question, given the vast amount of available water in our oceans. This, unfortunately, creates significant consequences. Initial costs can vary from US$30 million for tiny plants to over one billion dollars for major desalination structures. The amount of power required constitutes about half the total running expenses. Sea creatures die from being caught in screens or sucked into the plant. The process of desalination discharges brine and wastewater which can further harm nearby aquatic life.[15]

The most obvious part-solution to water shortages is to use less, which can reduce demand by up to 50%. The technology for recycling waste water is improving rapidly, and is less expensive than desalination. Enhanced storage of storm water run-off, whether in home water tanks or mini-dams, is another viable option. Individual attitudes matter the most. We need to value each tiny drop, given that every one of us is responsible for our daily use.

Turn off running taps, only use the dishwasher or washing machine when they are full, have short showers rather than baths. When a whole city is faced with seriously low dam levels in a drought, people rise to the challenge. Water consumption can often be more than halved when everyone makes a consistent effort to lower their usage.

Issues: Entire countries are loathe to minimise their water consumption when they are upriver, but they need to accept that everyone downstream needs water too. Climate change reduces glacier sizes, which lowers the amount of melting ice available downriver. Difficult decisions on water allocation will need to be made and enforced. Individual 30-min showers are OUT.

Probability rating of most people consistently eating insects by 2075: 90%

Probability rating of lowering food waste from 30% to 15% by 2075: 80%

Probability rating of humanity cutting personal water use by 50% by 2075: 80%

PROVOCATION FOUR. The Reimagining of Work

Office employees must be getting bored with age-old titles such as secretary, or head of human resources. More recently, some creative descriptors have begun to appear on company websites. What about Digital Overlord (Website Manager), Wizard of Light Bulb Moments (Marketing Director), Sales Ninja (Sales Executive), or even Dream Alchemist (Head of Creative)?[16]

Those titles already exist. But what about entirely new occupations that will appear in the future? Perhaps you might become a 100-year Counsellor (helping people to live a fulfilling life as they age), Swarm Artist (controlling drone swarms to create art or music experiences), Nostalgist (supporting the elderly to recreate remembered experiences), or possibly a Genetics Coach (helping people to understand their genetic profiles and the impact of genes on their health and wellness).[17]

Regardless of what work we do, what will drive our enthusiasm for it? Money is obviously one factor, although younger people consistently express their desire for work that is socially just, fulfilling, and that challenges their thinking. They will engage in many forms of employment during their lifetime, given that they generally recoil from the thought of remaining in a single work position for forty or more years. Some have watched their parents endure long work hours over many years in an unloved job, and the prospect does not interest them in the least. They want work to be an inspiration and a core component of a quality life.

How do we best prepare young people for a volatile work world up ahead? One good start would be to stop telling them horror stories about life in their future years. That futuristic world will have plenty of issues, yet it also might be the most inspiring time in human history. Instead of frightening them, prepare our youth by helping them to develop the capabilities they will need in that future.

A capability is something that enables them to deal more competently with whatever happens in the rest of their life. These capabilities include critical thinking, initiative, creativity, empathy for others, resilience, and work agility. These are capabilities that robots may not be able to demonstrate for a while yet.

While AI is going to disrupt most workplaces, it also will rescue us from the tedious components of work. Think of earlier technologies that saved us time and energy. Washing machines, word processors and electric tools have saved hours of labour. In past industrial revolutions, the initial consensus was that the new technologies were going to replace workers. On the contrary, the technology created more jobs and improved productivity more than ever before.

What everyday form will this work take? The five-day work week is under threat in many business settings, with claims by workers that equivalent productivity can be provided in just four days. The COVID-19 era ushered in the very necessary WFH (Work From Home) process during lockdowns, with employees discovering that they quite liked avoiding rush hour traffic and office politics.

There are other work cycles becoming more evident. US gerontologist Ken Dychtwald, co-author of *Age Wave*[18], suggests that many of today's under-30s will live in flexible seven-year work cycles, rather than a linear one-job, 50-year pattern. For six and a half years, they will learn and earn in a specific profession, and for six months will 'retire' to pursue a lifetime ambition.

This might involve the study of horse breeding in Mongolia or learning how to become a whirling dervish in Türkiye. Some may take seven years to complete this cycle, some four years, but the principle stands. Given the difficulty of governments affording pensions in the future, they also may continue with this 7-year work pattern until they are 92 years old.[6]

We might even enter a work-less society and be partly paid with some form of a UBI (Universal Basic Income). This already occurred during COVID-19. Perhaps a robot (built by another robot) or our online digital twin will perform some of our work for us, and we will receive the income for that techno-effort. Maybe full-time work is not meant to stress us out and contribute to dysfunctionality in our families. Instead, part-time work might be viewed as a worthwhile and rewarding part of being alive.

Aristotle talked of *eudaimonia*, a Greek word that translates as 'human flour-ishing'. This flourishing is derived from experiences that fulfil the soul and focus on the human good, steadily enabling you to become a decent human being. Perhaps *eudaimonia* will become the trend of the 2050s.[6]

This is not an argument for laziness. It is a challenge to reconsider what it means to engage in work that has depth and purpose, rather than one where the struggle to earn an income overwhelms the joy of being alive. Paid work can fill up more than one-third of the hours in our lives. Surely we want our children to free themselves from the trap of endless soul-destroying work during those hours.

Issues: History might repeat itself, and we may see more jobs being created than ever before. However, the transitioning into those new jobs may be too onerous for some, given the amount of retraining that will be required. We also need to pay deep respect to the development of human-centered workplaces, rather than placing emphasis upon artificial online entities or humanoids overwhelming our real life.

Probability rating of everyone being replaced by robots at work by 2075: 10%

Probability rating of AI performing most of the tedious parts of your job by 2075: 95%

Probability of an astonishing new array of occupations by 2075: 100%

Did you notice the occasional optimistic perspective in all these Rubberband Effect provocations? Did you also take note that little was mentioned about nuclear warfare, or virulent pandemics, or dramatic climate change if we fail to take strong action? There is always a percentage probability for all of these. 1%, 25% and 60% by 2075 AD if you would really like to know.

The reality is that the planet will always need to cope with the good and the bad. There will continue to be existential crises that might profoundly threaten some of our lives. At the same time, we will be constantly astounded by the initiative and resilience of billions of our fellow citizens.

It's not easy to deal with exponential change. The human brain is hard-wired for steady and incremental refinement. Some of us would just prefer everything to stop for a while, which may have some merit at times. These four provocations were purposely designed to stretch your thinking well beyond what you may usually ponder. One thing we can predict with near-certainty in the future is that the rate of change will be much more rapid. You will need to take that rapidity into account when you predict anything. What previously took twenty years to eventuate may soon take two or less.

Predictive Exercises to Expand Your Predictive Ability

- Brainstorm 10 highly improbable business ideas for 2050. Then, refine these ideas into more realistic and actionable business plans for right now.

- Explore Moore's Law and project its impact on a specific technology (e.g., computing power, data storage, AI capabilities) up to the year 2050. Predict how these advancements might influence daily life, work, and society.

- Read articles that predict what sort of work will be done in more than 20 years from now. Based upon what you read, make your own predictions, and develop a % probability for each.

- Search for 'Future Predictions' online (there will be plenty), and add your own addition to each set of predictions you read.

- Research current advancements aimed at increasing human lifespan (e.g., gene editing, nanotechnology). Predict the societal, economic, and ethical impacts if the average lifespan were to reach 150 years by 2075.

- Write a list of future events that would be impossible (according to today's standards). Then conjecture on a probability rating and a time for the introduction of each of those events.

- Reflect on Aristotle's concept of eudaimonia (human flourishing) and how it can be applied to future work environments. Predict changes in societal values and work structures that prioritize human flourishing. Design a prototype of a future workplace that fosters eudaimonia.

CHAPTER 8 IN SUMMARY

- The concept of exponential change, exemplified by Moore's Law, suggests rapid advancements in technology and various fields, making future predictions increasingly complex. Understanding this pace is crucial for predicting future developments.

- Imminent technological advancements include brain-computer interfaces, AI integration, and genetic engineering. These could significantly enhance human capabilities, though ethical and social implications must be carefully considered.

- Advancements in healthcare and technology could dramatically increase human lifespan, possibly even leading to some form of artificial immortality. This raises questions about societal impacts, resource allocation, and the quality of extended life.

- Addressing future food and water security involves innovative solutions such as lab-grown meat, insect-based diets, and improved water conservation methods. Global cooperation and individual responsibility are key to managing these resources sustainably.

- The nature of work is expected to evolve with new occupations and flexible work patterns. Emphasis on human capabilities like critical thinking and creativity will be vital, alongside AI handling more repetitive tasks. Societal values may shift towards work that promotes human flourishing and well-being

CHAPTER 9: DEVELOPING YOUR OWN PREDICTIVE FUTURE

If you don't know where you're going,
any road will do.

LEWIS CARROLL

The Psychology of Prediction

If you have ever used a streaming service such as Spotify or Apple Music to create a playlist based upon your mood - energetic, creative, meditative - then you have engaged in what is called affective prediction. The service considers your previous favourite music, and generates a list that is more likely to create your chosen mood. In your own life, affective prediction - or affective forecasting - is when you predict how you might feel at specific times in the future.[1]

Predictions or decisions about a future event can be influenced by your present mood. The dilemma is that those choices might be made when you are in a high energy state, and yet the event itself might take place when you (and other attendees) are low in energy. There is a marked energy difference between Mondays and Fridays for most people. Even in a single day, you can experience high and low energy times. When you predict or plan an event, do not be influenced by your vitality at the present time. You are more likely to set yourself up for failure. Think of how most people – including you - will feel during the event itself.

We underestimate the degree to which our emotional states affect our ability to do what we intended.[2] When we think ahead to a critical conversation planned with a partner later in the night, we do not always account for our lowered emotional tiredness at that time. Instead, our planning is often based upon how we feel right now. We may also not account for the lowered energy levels of the other person in the conversation. Plan your pending discussion with care. It is not always easy to consciously think ahead, especially with predicting your emotional state at that time.

The reality of a busy life is that we prioritise our present time over our future time. Time management expert Laura Vanderkam describes three ways that we experience our lives. The 'Remembering Self' is what happened in the past, the 'Experiencing Self' is what happens for you right now, and the 'Anticipating Self' plans and looks forward to events in the future. She refers to this as "our three selves".[1] According to Vanderkam, we too often pamper the present, and fail to account for the emotional impact on the future self. This Anticipating Self will not be grateful for that overload up ahead.

It's All in Your Biased Mind

Whether you think about the past, the present, or the future, your thoughts are influenced by your cognitive biases. Very little in life is black and white, given that your biases colour everything you see and experience.[3] Here is an example. Your boss has just announced the winner of a company award (and it's not you). How does your brain react to this news? Do you think "Hard to believe. She's not the best worker around here" (Negativity bias), or do you ponder "Good luck to her. She treats everyone well" (Positivity bias)? What about "She's too old to manage her work anyway" (Age bias), or "People from overseas win too many of these awards" (Racial bias)?

You can experience biases in terms of context, overconfidence, and mood. As well, dichotomous thinking - or black and white thinking - can too easily lead to you ignoring information that may have been highly beneficial. It's a wonder that you can ever be completely objective with anything that you think or do. The likelihood is that you may not. Your brain is a maelstrom of biases all through the day, and these biases can be fuelled by your lifelong stream of thoughts and experiences.

The concern with your biases is not just how they affect your thinking and emotions right now. They also have influence on your predicting and planning for life up ahead. As you ponder the future event, you might find yourself experiencing thoughts such as: "This group of oldies never get what I'm talking about", or "Planning by the interns is a nightmare. Something always goes wrong". These are a combination of specific biases, allied with an overall negativity bias, which is one of the most pernicious of all. Whether or not you might be depressed, the negativity bias registers more strongly in your brain than the positivity bias.[4] This partly explains why many people predict that the world up ahead will be negative.

Admittedly, there can be merit to the negativity bias. Thousands of years ago, this bias helped to keep us alive. While hunting, we needed to be aware of the dangers around us. Even an everyday greeting to a stranger may have had a deadly ending, and we would have been constantly on guard against a dangerous situation. These events, and others that led to excitement or anxiety, created adrenaline in our system which heightened awareness of the experience and hopefully kept us alive.

When someone today perceives that their life is lacking in adventure, they sub-consciously seek an adrenalin surge from the negativity they are encountering. Modern media have taken advantage of this brain functioning. Negative headlines on social media are more likely to be opened and read than positive ones.[5] The first report on the evening news is invariably biased towards doom and gloom.

You will reside somewhere along the negative to positive spectrum according to different influences in your life. Beware the extreme ends of the line. Total negativity is a serious concern, and professional support may be worthwhile for you. Being overtly positive can be just as much of an issue. You will believe everything you hear, and will be susceptible to scams and trickery. A better option is to hover at the 90% mark on the line towards optimism. Stay open to opportunity, but maintain vigilance against the occasional fraud.

The self-fulfilling prophecy has been studied exhaustively. The research generally demonstrates that something in your life is more likely to happen when you believe that it will.[6] Being convinced that you will fail an exam too often leads to a substandard result. To reinforce on your prediction, you may not study effectively, or you will question whether you are smart enough for the subject.

Even worse, you might place your faith in some vacuous positive thinking that is not based upon reality. Don't Worry Be Happy is not good advice. You may also be influenced by the self-fulfilling prophecy that others hold about your ability. If you are surrounded by lecturers or work cohorts who expect the worst from you, it can compromise your performance.[7]

Your biased brain can even be tricked into believing something that is not occurring. The placebo effect - such as being given a sugar pill rather than proven medication - can sometimes help to improve some symptoms. Conversely, the nocebo effect is when a person's negative expectations about a treatment lead to the end results being even worse than they otherwise would have been.

Your forecasting biases can affect happiness levels.[2] A strong focus on the negativity bias can lead to a greater chance of a poor future performance, which impacts upon your self-worth. A study from the University of Bath School of Management in 2023 also found that some women are less willing to take risks than men. Their perceived emotional pain from future losses is greater than the positive benefits that might result.[8]

What Makes you Better at Prediction

Becoming aware of your biases can enhance your predictive ability. What else can make you a better forecaster? Let's explore some brain-based indicators:

YOU ARE A CRITICAL THINKER. You step back and become aware of your own thought processes. You realise that your thinking needs to be analytical and thorough. You ask yourself questions such as Q. What is influencing my thinking with this topic? Q. Why am I so uneasy about my initial projections? You check constantly for mistakes you may have made, and you think in terms of uncertainty. Rethinking what you have predicted is a regular occurrence.

YOU DEMONSTRATE A GROWTH MINDSET.[9] You love learning, and you savour discovering different perspectives. Conversely, in a fixed mindset, you would curtail any decent efforts to learning something new, and you would rather not read an article that is opposed to what you believe. There are some indications that the older you are, the less likely you are to change your mind. Philip Tetlock explains that most experts are "too quick to make up their minds and too slow to change them."[10] Society needs to value people who adjust their mindset, rather than those who stubbornly hold on to age-old and disproven beliefs.

YOU FOCUS ON PRECISION. Anyone can make a vague statement about something occurring in the future e.g. tech stock will go up over the next two years. Good predictors, such as you, will be much more precise about your projection, and will do the reading necessary to calculate a clear projection into the future. For example, you may research and then arrive at the probability that "Nvidia shares will rise by 4% to 5% within the next 8 months."

Your affective attributes – derived from your emotional intellect - are equally critical when you want to enhance your predictive ability. Here are two that can benefit you.

YOU ARE A TEAM PLAYER. There is merit in sometimes working with a diverse team. Other people can offer different perspectives that did not occur to you. Their musings might spark an idea that leads to a much more powerful prediction. The one issue? Be aware of what is called group think, in which you all agree on everything because you do not want to appear rude. Instead, each team member must respectfully challenge ideas when warranted, and each must accept the response to that challenge without becoming defensive.

YOU EXPRESS HUMILITY. You know that few predictions will ever be perfect. No matter how exhaustively you conduct your research, there is still some likelihood that you might be incorrect. It is advisable to approach all predicting efforts with deep humility. Refrain from overconfidence, and accept that you will sometimes be incorrect. Be prepared to change your mind, be endlessly cautious with your calculations, and avoid black-or-white thinking.

Predicting At Work

Predictive ability enhances your worth for your employer. Being able to accurately measure economic or social trends is invaluable to any business or public service, and an employee skilled in predictive analysis is highly coveted. Such ability can improve your marketing acumen, it can enhance your future planning ability, and it can even help you to detect fraud.

Predictive models - such as neural networks, time series data mining, decision trees - are used in many professions.[11,12] Those professions include insurance, marketing, healthcare, finance, and weather forecasting. Banks use the models to detect fraudulent activities. Hospitals determine which procedures lead to optimum patient outcomes. Manufacturing companies forecast the probability of machinery breakdowns.

Here are four predictive modelling processes used in the workplace:

Linear progression introduces a single variable to your future projection. If you were graphing a trend line on expected profits over the next six months with your online clothing company, the one extra variable might be the specific promotions that you have undertaken. Other individual variables might include daily weather conditions, or a recent alteration to your company branding.

The Delphi Technique[13] taps into the collective thinking of a group of experts. Each person provides his or her answer to several rounds of questions, and the responses are shared anonymously with the team after each round. The questions might focus on future marketing options, or possible improvements resulting from the introduction of a new technology.

Scenario planning is a strategy for mapping a range of future options.[14] If you were thinking of starting a small business, you might place four headings along a line, namely: Worst, Passable, Good, Best. Under each heading, you would describe what that scenario would look like if you proceeded with this startup. If you work with a major corporation, they very likely invest significant time and money into developing worst to best case scenarios for their specific product or service.

Question / Research / Predict / Analyse. Many predictive processes align with these four stages. Here is an example of placing them into action.

1. **Question**: You clarify a key question or questions about what you want to predict. For example: To restock your online store, you might ask yourself: what specific items of clothing are most likely to be in future fashion? What 'look' will be in vogue in 8 months from today?

2. **Research**: You find as much contextual data as possible. You conduct a search on recent fashion shows and read everything that five of your favourite influencers have written in the past few weeks.

3. **Predict**: You make an interim prediction and assess the percentage probability of it taking place. The past year has featured a formal style, but your research indicates that 'casual denim' is about to come back in. You forecast an 80% chance of this becoming commonplace in 8 months from now.

4. **Refine**: You then spend several days on revisiting the prediction. Some further data indicates that the casual will be integrated with a sporty flair. You stay with your 80% prediction and add that a sporty look will be a key feature.

There are many variations on these processes, and you need to find what works for you. One extra worthwhile step is to analyse your success (or otherwise) after the predicted event has taken place. This would be your own version of hindcasting. Ask yourself: Q.1. What worked in this whole process? Q.2. What didn't, and why? From there, build those learnings into the next predictive analysis that you do.

Predicting in your Personal Life

Make predicting your new personal superpower. While others may want to fly or use X-ray vision, your special gift may be forecasting most of what will happen in your future. This is not some sort of magic act that you perform at your next party. It simply requires you to humbly forecast most of what will occur up ahead.

How can you build some prediction models into your everyday life? Start with 'future projection', which is a form of mental time travel or self-fulfilling prophecy. You may already do it sub-consciously every day, although awareness of the process improves its application. Future projection is a ten second brain video of a specific future event. If you can, visualise it in full colour, and hear the sounds, and sense how you are feeling during the experience. Look at this through the eyes of the camera operator, and see yourself as though you are featured on a large screen. Play this brain video at normal life speed.

Here is one practice option for using this future projection. Think of an activity you are likely to do in the next few days for each of these four categories: exercise, nutrition, mental wellbeing, and relationships with loved ones. Perhaps you ended up with activities such as a 30-min jog late on Monday afternoon; vegetable stir-fry for dinner on Tues night; listening to meditative music for the last hour before Wednesday night bedtime; playing basketball with your son late on Thursday afternoon. Visualise a brain video of each activity. Practise this future projection process until it becomes part of your everyday thinking. Take note that energy follows intention.

If you paid attention during your high school science lessons, you may be familiar with the Scientific Inquiry model.[15] Its beginnings formed in Socrates' mind, and the steps are more of a cycle than a straight line. This is a variation on the Question/ Research / Predict / Refine process. This model requires five steps, namely:

1. Make an observation. "I can't turn the TV on."

2. Ask a question. "Why won't it turn on?"

3. Form a hypothesis, or a testable explanation. "Maybe the batteries are flat in the remote control."

4. Make a prediction based on the tested explanation. "If I replace the batteries, then I might be able to turn the TV on."

5. Test the prediction. "If the remote control works, then my hypothesis is correct. If it doesn't work, then I need to go back to Step 3."

When you persevere with this inquiry process for a few weeks, you will eventually find yourself performing it almost sub-consciously with issues in your life. It can merge with the future projection process, with you running your brain videos for each of the steps. Teach your children this process the next time they are complaining about an issue in their lives, and their problem-finding and problem-solving skills are likely to improve. This applies to children of all ages, including you.

Driving a car is analogous to these predictive processes. You do not watch for what is up ahead by using the rear-view mirror. That is your past, not your future. To make progress on the road, you look out through the windscreen, and constantly predict what is up ahead. I once received some advanced driving training, and was given one piece of advice that I have always remembered. The instructor said: "Look as far ahead as possible, and you will have a better chance of predicting any issues." My driving improved, and my attitude to looking ahead in my life received a boost.

Predictive Exercises to Expand Your Predictive Ability

- Set a personal goal for the next six months. Predict the challenges you will face and the milestones you will achieve along the way. Track your progress and refine your predictions over time.

- Practice 'future projection' by visualizing a specific upcoming event (e.g., a presentation, a family gathering) in vivid detail. Include sensory experiences (sights, sounds, emotions) and play the scenario at normal life speed.

- Identify an upcoming challenge (e.g., an exam, a work project) and create a positive prediction statement (e.g., "I will excel in this project because I have prepared thoroughly"). Write down steps you will take to reinforce this positive outcome. Reflect on how your actions and mindset align with your prediction.

- Consider current trends in remote work, automation, and AI. Predict new job roles that might emerge in five years from now, and the skills that will be in demand. Develop a personal plan to acquire some of these skills and for you to adapt to future job markets.

- Use the scientific inquiry model. Generate a key question about an issue, and then develop a hypothesis that can resolve the issue.

- Predict how you will feel during an event up ahead. It's called affective forecasting. Learn how to prepare for those feelings eg being nervous.

- Predict something that is likely to happen to you when you are with a specific group of people. As you engage in the predicting, clarify which biases - negativity, age, positivity - you might be applying.

- The 'wisdom of crowds' has been well-authenticated for predicting the future. Find a small group of friends / colleagues and experiment by using the Delphi Technique.

- Make a detailed prediction about a personal or professional goal (e.g., fitness achievements, business growth). Revisit this prediction monthly, assessing its accuracy and adjusting as new information becomes available. Emphasize humility in your approach, acknowledging uncertainties and learning from inaccuracies.

CHAPTER 9 IN SUMMARY

- Understanding how emotional states influence future planning is crucial. Affective forecasting involves predicting future emotions and can help in making better decisions by considering how one might feel during the actual event, rather than relying on current emotions.

- Cognitive biases, such as negativity bias and overconfidence, affect predictions and planning. Recognizing and mitigating these biases can lead to more accurate forecasting and better decision-making.

- Effective predictors are critical thinkers, demonstrate a growth mindset, focus on precision, work well in teams, and express humility. These attributes enhance predictive accuracy and adaptability.

- Predictive models, such as linear progression, the Delphi Technique, and scenario planning, are valuable tools in various professions. These models help in making informed decisions, forecasting trends, and improving future planning.

- Developing personal predictive abilities involves practices like future projection, the scientific inquiry model, and setting specific goals. These techniques help individuals better anticipate future events and align their actions with desired outcomes.

CHAPTER 10: DEVELOPING THE PLANET'S PREDICTIVE FUTURE

*We are like butterflies who flutter for a day
and think it is forever.*

CARL SAGAN

The Short-Termism Dilemma

One of the oldest democracies in the world, the Haudenosaunee Confederacy (Iroquois), espouse a philosophy called Seven Generations.[1] It means that you pay respect to the people who will follow you in the next seven generations (about 150 years ahead). An alternative timeline places you in the middle of the seven generations, with three before you, and three after you. The ones preceding you are your great-grandparents, your grandparents, and your parents. In the future, the next three generations will be your children, your grandchildren, and your great-grandchildren. According to this philosophy, you learn from your elders, and you then pass those teachings on to your descendants.

The dilemma is that many of us today only focus on the short-termism of our own lifetime.[2] The relentless rush of everyday life can narrow our perspective. Just getting through a week feels overwhelming. How can we think about the far future when reaching your next holidays feels like a massive achievement? Most modern societies do not easily ascribe to a long-term vision for their complex collective lives. To ask what life will be like in 2100 AD usually elicits a response such as: "You can't possibly predict that far ahead. And you certainly can't influence it."

Conditioned into thinking about our own life span, we predict and plan accordingly. We ask: What work will I perform in my 40s? How will I best raise my children? What sort of retirement am I seeking? How do I plan for that? Our short-termism is reinforced in myriad ways. We experience

3-to-5-year political cycles, allied with the governmental policy changes that accompany those cycles. We hope for a quick-fix financial gain on the stock market or in the Lotto. Our businesses work in quarterly cycles, and some social media apps compress video viewing time to a scant six seconds or less. We have become so centred in the near-future of a minute, a day, or a week, that we have lost our collective patience on preparing for the generations ahead of us.

The political and even ethical dilemma is that, given the limited funding, societies are forced to choose between resolving critical issues right now, or minimising the impact of those issues on future generations. Climate change is probably the specific concern most relevant to this thinking. Do we protect the livelihood of millions of workers today; or partly sacrifice those jobs to create economies that support a viable future environment?

If it needs to become a population popularity contest, then the numbers favour choosing the future. About one hundred billion people have been alive in the past fifty thousand years. Today our population is above eight billion. If we project into the next fifty thousand years, the number of people alive during that time period may be up to 6.75 trillion.[2] This projection is based upon present birth rates, which is unlikely. Even if the number is 'only' one trillion, we need to pay respect to them.

Predicting the future is too often based upon what will happen soon. Global media commentators focus on one to five year futures. That short-termism may need to change. What about 2135 and 2150 AD? For you in your everyday life, perhaps the most powerful approach is to base your contributions on how they will impact on future citizens not yet born. Chief Seattle was quoted as saying: "We do not inherit the planet from our parents, we borrow it from our children."

Doomsday Predictions

When it comes to future predictions, one group of scientists is not very confident. It is 90 seconds to midnight right now, according to the Doomsday Clock, featured in the *Bulletin of the Atomic Scientists*.[3] First introduced in 1947, this symbolic clock is the opinion of a group of eminent scientists on how near the world is hypothetically to nuclear disaster. In 2007, its brief was expanded to include climate change. The time setting has varied over the years, from an optimistic 17 minutes to midnight in 1991 to the present more pessimistic evaluation.[4]

Over the past two thousand years, the planet has been inundated with end-of-the-world predictions. These narrative have featured fire, pestilence, earthquakes, and even the trailing dust from a passing comet. Doomsayers have several points in common. Until now, few have been factually correct. Another is that, when their predicted date of total destruction passes without incident, they admit to a mathematical error, and reissue another date for the pending doomsday.

What drives the delivery of these prophecies? Some people's negativity biases dominate their life. Embittered citizens perhaps want some notoriety. Business people see an opportunity to sell products to people who are easily frightened. Politicians may need a political distraction from other crises, or seek to incite hatred for specific groups. Sometimes, a cult leader is simply deranged.[5]

Some religious elders in past times reinforced their power by creating fear amongst their followers. This included threats of torture or even death for non-believers. Most religions provide a prophecy that the world will come to an end sometime. The concept of 'end' is open to interpretation. It does not necessarily mean that the planet will be physically destroyed. It might refer to a dramatic change in sociological circumstances. Like most other horoscopic attempts at predicting a destructive future, the wording is vague enough that anyone can reframe it to mean almost anything.

There are creative variations on how these religious cataclysms might occur. The Mexican Aztecs between the 14th and the 16th centuries asserted that a never-ending solar eclipse would destroy the world. Buddhists predict that after we have witnessed a comprehensive degeneration of human society, the planet will attain a new era in which the next Buddha will appear. Zoroastrians are taught that the planet will be consumed by fire, and that all sinners will be punished for three days -- and thereafter forgiven.

The Christian bible maintains that the world's end will be signalled by a breakdown of society, which will feature calamity and war, and the righteous will be raised straight to heaven. Hindu practice does not actually predict an end to the world, and this is because they believe that the universe traverses a sequence of cyclical eras.[6]

Critics of religious practice often point out this negative endgame narrative when they are decrying these faiths. They contend this is hardly the future we want young people to ponder. To be fair to the religious faiths, many of them also espouse core sets of values that provide guidance for a fulfilling life. In their purest form, most religious practices encourage unconditional love and deep connection between their followers. This focus on empathic communication is a much more optimistic message to offer to our youth.

Other predictive doomsday analyses are not so easily dismissed. The Club of Rome, a think-tank of diplomats, scientists, and former heads-of-state, have provided continuous compelling data about the state of the world over several decades. In 1972, its first report was developed by a scientific team from the Massachusetts Institute for Technology (MIT). This report - The Limits to Growth - was based upon a computer simulation. Variables such as population growth were constantly readjusted, and the simulation was run over and over again.

The report indicated an eventual clash between the advancement of industrial civilization and the finite availability of resources. According to the report, stressors would develop by 2030, and reach a tipping point around 2040. More recent data analysis in 2021 has confirmed that the 1972 report's findings remain robust.[7] These predictions are not inevitable, and may become altered by future technologies or political choices, amongst other effects.

Many civilizations have risen and fallen in the past six thousand years. Very few have lasted more than several hundred years. It is part of the cycle of birth, growth, and usually, some form of collapse. The possible causes for that collapse can include mass migration, pestilence, war, and natural catastrophes. While some of these causes are still possible, we can now include others on the list such as nuclear warfare, robot revolutions, and the uncertain implications of AI progress. Is our present civilization in trouble? In its present state, it is likely to experience a restructure.

Some civilizations have regenerated, such as Egypt, India, and China, while others such as the Roman Empires and the Maya never do.[8] History might repeat itself, although one debilitating factor today is the persistent negativity bias pervading our airwaves and everyday conversation. This bias reinforces a belief that life is not as good as it could be. In spite of indisputable long-term data about global improvements in our standard of living, a paradoxical situation arises. The better that things are, the worse that people believe them to be.

The Paradox of Progress

A paradox is a statement that seemingly contradicts itself. In his 1962 novel Catch-22, Joseph Heller described the paradox of a soldier who wanted to avoid combat by being declared insane. He was eventually declared to not be insane because he was sane enough to understand that insanity could relieve him of combat duties. In everyday parlance, a Catch-22 descriptor is attributed to a no-win situation e.g. You can't get a job without experience, but you can't develop that experience without a job.[9]

The paradox of progress explains that the better that things get, the worse we think they are. This formed the basis for *The Progress Paradox* by Gregg Easterbrook[10], in which he documented two competing influences. One is the unrelenting improvement in many worldwide indicators of life quality. The second is our steadily diminishing level of happiness and perceived wellbeing over the past seven decades. This in part is due to the endless media diatribe that focuses on what is wrong with the world.

In spite of the many improvements to our standard of living, too many of us are choosing — or have been convinced — to believe that the world is not a good place, which results in a lowered confidence about the future.[11] This impacts upon any predictions we might make about the world up ahead.

Are there serious issues out there? Of course there are. You do not need to be a predictive genius to know that some dilemmas could badly compromise the quality of life on the planet. For the first time in human history, today's youth may not experience a better standard of living than their parents. There is credence to this belief, especially when house ownership is considered. In a just society, surely everyone deserves to have a roof over their head every night.[12]

There is some light at the end of these very long tunnels. In spite of - and sometimes because of - those dilemmas, millions of people constantly engage in actions that directly address these problems. The collective human efforts of the past two hundred years have created improvements to global life that are simply astonishing.

If you think that life today is getting much worse, then compare it to times since the early 1800s. 90% of the world's population back then lived in poverty. Two-thirds lived in extreme poverty in 1950. In 1981 it was still 42%. Today it is less than 10%, although that depends on where you draw the financial line in the sand with calculating global poverty. In 1820, adult literacy rates were around 10%. In the 1930s, this had risen to one-third. Today it is over 86%. High literacy rates are a strong indicator of a decent quality of life.[13]

In the early 1800s, around 43% of the planet's newborns died before their 5th birthday. By 2017, child mortality had dropped to 3.9%. This is due to improved health standards, including immunization, better nutrition, and enhanced hygiene.[13] Will all these issues improve further? That will depend upon a range of factors such as the burgeoning cost of health care. However, if historical trends are any indication, improvement is likely.

It does not matter what indisputable data is explained to some people. They will remain convinced that the world is growing increasingly dangerous and dystopian. Awareness of positive directions does not necessarily create better mental wellbeing, although this respectfully will depend upon your standard of living right now. If you are jobless and homeless, the world is unlikely to look like a decent place.

Surveys worldwide on 'happiness' and 'quality of life' offer mixed results, although they are predominantly in the negative. Depending upon how the questions are asked, somewhere between 40% and 75% of young people today are seriously concerned about the future.[14,15,16] The reasons for this dystopian perspective usually focus on job certainty, financial wellbeing, and climate change. In recent times, the circumstances around COVID-19 have exacerbated these negative beliefs. The propensity for worry is not just a modern affliction either. In the 1940s, people worried that the war would never end. In the 1970s, the concerns centred on global starvation. Today, it is whether the robots will take over our jobs and our lives.

If we time-travelled back 30 years, we would find that the global worries at that time rarely became the disasters we expected. We simply applied our human ingenuity and resourcefulness, and created solutions whenever possible. When today's youth look back three decades from now, will they also find that few of our gravest fears eventuated?

Our Talented Humanity

One powerful factor might partly put their mind at ease. That factor is the inspiring innovation, resilience, and adaptability that has been demonstrated time and again by human beings throughout history. This will likely continue to be displayed well into the future. Remember: this is the civilization that invented chocolate. That is enough to earn credit points just in itself. Added to that is the wheel, the needle, the printing press, penicillin, the telephone, the lightbulb, the internet, batteries, planes, refrigerators, vaccines, and X-rays.[17] Every one of these inventions resulted from an issue needing to be resolved. We are the solution masters of the universe.

The world has also created the most astonishing works of art, and music, and literature. We have created the Olympic Games, and in spite of the occasional boycotts, most countries on the planet join together in celebrating the event. Humanity has placed people on the moon, and is likely to send humans to Mars before 2040 AD. We built the Egyptian pyramids long before giant engineering machinery could lighten the workload.

We created the United Nations in 1945, and although it is subjected to periodic transgressions and politicking, it at least is a global body supported by nearly all nations. Where would we be without the genius of these contributions to our culture? You can reasonably predict that the collective desire and determination to create new contributions will continue unabated.

Humanity has equally demonstrated its capacity to recover from adversity. At the personal level, most of us have faced debilitating illness, or the loss of loved ones, or tragic accidents, or a breakdown in long-term relationships. Collectively, we have been subjected to major crises that required significant global resilience. World wars, pandemics, and major natural disasters have tested our physical, mental, and financial limits. The inspiring feature of nearly all disasters on this planet is the willingness of other countries to offer support.

In his heartening book called *Humankind: A Hopeful History*[18], Rutger Bregman wrote that "It's when crisis hits - when the bombs fall or the floodwaters rise - that we humans become our best selves." Yet we do not just exhibit our best selves during such disasters. It is becoming more evident in our lowered incidence of global violence.

In *The Better Angels of Our Nature*, Steven Pinker wrote of a measurable decline in violence - evidenced by a humanitarian revolution - through the past few hundred years. In the 17th and 18th centuries, we experienced lower trends in the use of judicial torture, slavery, sadistic punishments, and even superstitious killings. Referring to them as 'better angels', Pinker outlines four motives that have led us away from violence and towards altruism and cooperation. One of these motives is empathy, which is the ability to identify with the feelings of another person. This 'better angel' has every chance of continuing to fly, and was evidenced by the extraordinary support given worldwide to countries that were ravaged by COVID-19.[11]

Millions engage in crowdsourcing initiatives, and support each other with positive political movements. The X-Prize[19] is a global platform for leveraging the positive power of world-wide competition, and is supported by wealthy individuals who donate money as prizes for resolving massive issues. Its specific domains include biodiversity and conservation, climate and energy, health, education and deep space.

Their largest ever competition, worth US$101 million and launched in November 2023, will span seven years and incentivise medical teams to test and develop therapeutics that enhance healthy aging. Exciting projects such as these are the narrative we need to share with each other about the planet's possible future. When we predict and create what is up ahead, it is worth accounting for the success stories that humanity has already demonstrated, and that it will continue to do.

The Global Self-Fulfilling Prophecy

The daily stories you see and hear shape your thinking about the world up ahead. Subjecting yourself to a tirade of horrendous news leaves you fearful, and expecting the worst. You become what you consume. This refers to both your body and your brain. If you share bad news with family, friends and colleagues, the unsettling messages spread further. Conversely, engaging with a good news story generates hope in you and others, and can restore your faith in humanity.

Consider this in terms of the billions of people who are exposed to mass media every day. The reporting can lead to the reality. Rumours about the fragile viability of a bank can panic customers into withdrawing their savings, which can lead to the bank collapsing. Media reports about pending food or fuel shortages often lead to a real shortage, as people rush to buy the product. On a more optimistic note, media coverage of an innovative local charity that supports homeless people can lead to significant financial support. The charity then instigates further help for the homeless. Initial reports end up creating the reality.

The self-fulfilling prophecy is derived from the Greek myth of a sculptor called Pygmalion who so profoundly loved the statue of a women he created that she was brought to life. Because of his adoration, she married him, thus creating a self-fulfilling prophecy. If we could all only see the Earth from outer space, it would possibly evoke a similar feeling. Is this ever going to really happen with over eight billion people? Given the complexity and diversity of the human race, it sounds fanciful, although time and again, humanity has proved that it can do whatever it takes to redress an issue.

When I describe our many astonishing human achievements, people invariably respond with "but they're global leaders or very wealthy people. I could never do something like that." Granted that there are some outstanding humans who never seem to sleep, although that generates an assumption that the rest of us contribute little to the planet. Let's move on from that belief.

You add to the planet's welfare when you raise your own kids. You contribute with your paid and volunteer work every day. Any work that is legal and ethical provides benefit to your community. You improve the world when you support a charity, either financially or practically. The sum total of everyone's efforts create the planet's future.

Sadly, if we are to believe in the self-fulfilling prophecy, then the near-future may not be the best time in human history. Many postulate on the inevitability of a pending dystopian era. What can be done to reverse this rampant negativity in the mass media, or in the billions who are convinced that the future will not be positive? Read optimistic news sources that detail what is essentially good. Tell our children about what can be accomplished, rather than instilling fear by recounting what is bad. That fear robs us of the energy to create inspiring action. This does not mean we ignore what needs to be resolved. Rather, it fuels impetus to resolve those issues.

The quandary with the hundreds of years of steady global progress is that it is incremental. The changes cannot easily be seen every day. It's like watching children grow up. Millions of supportive actions contribute to their long-term development. Metaphorically, the world is a complex statue that is crafted with countless tiny chips into the marble. One chip seems to make no difference, although eventually, a beautiful statue has been created. It takes time. And when it comes to the global project in which we're all involved - the one where we predict and then create a sustainable future - we need to envision a viable world, and then chip away until it slowly takes shape.

This book has focused on predicting the future. The most powerful way of doing this is by believing that something is possible... and then setting out to make it happen. Improvements do not just happen at us, or to us. They occur because we decide what the future can be, and then we work on the implementation. In light of that, I predict the future will be a 'beautiful mess'.

In the mess of political dysfunction, and pandemics, and unsettling realities of climate change, we will still revel in the beauty of a child laughing, or a glorious sunset, or an inspiring piece of music, or of someone who creates a water purifier for poor villagers. What will be the future balance between the 'beautiful' and the 'mess'? That future will be whatever we collectively think it is going to be. Belief creates prediction creates action. So... what do you predict?

Predictive Exercises to Expand Your Predictive Ability

- Hold a debate on the merits and drawbacks of short-termism (3 years) versus long-term planning (more than 20 years) in political and business environments. Consider the impacts on social, economic, and environmental aspects.

- Simulate the Doomsday Clock by assigning time values based on current global threats such as nuclear proliferation, climate change, and pandemics. Propose actions that can move the clock back into a safer zone.

- Think seven generations ahead for your family. Yes, seven generations. Now predict some of the characteristics of that future family group that you could positively influence during your own lifetime.

- Analyze how media narratives shape public perception and create self-fulfilling prophecies. Track a current event and its media coverage, predicting potential outcomes based on the reporting.

- Start with a desired future outcome (e.g., a world with clean energy) and work backward to identify the steps needed to achieve that outcome. This helps in understanding the causal links and necessary actions.

- Ponder an example (if you can) of the progress paradox in your life ie the better that things have been getting, the worse you think they are.

- Develop a personal action plan for contributing to a sustainable future. Outline steps you can take in your daily life, community involvement, and professional endeavours to support long-term global well-being.

CHAPTER 10 IN SUMMARY

- The Seven Generations philosophy emphasizes the impact of our actions on future generations. This contrasts with modern short-termism, and highlights the need for a long-term vision to create a sustainable and thriving future.

- Historical and contemporary doomsday predictions often arise from negativity biases, political motives, or social manipulation. While some predictions are baseless, others, like those from the Club of Rome, are based on verifiable data and merit serious consideration to prevent potential crises.

- Despite significant improvements in global living standards, many people perceive the world as becoming worse due to constant negative media narratives. This paradox of progress highlights the importance of recognizing and communicating positive developments to foster a more optimistic future outlook.

- Human ingenuity has led to countless innovations and advancements, from the wheel to modern medicine. This capacity for creativity and problem-solving suggests that humanity can continue to overcome future challenges and create a better world.

- Optimistic thinking and actions can lead to self-fulfilling prophecies that improve the world. By envisioning and working towards a desirable future, individuals and communities can collectively shape a thriving global future despite the challenges faced.

THE THREE CORE QUESTIONS

If you have read this book with patience and thoughtful reflection, you may already be aware of the answers to the three core questions. Just to reinforce them with you, here are some responses:

Q. How can you really predict the future?

Chance events - the Lotto for example – cannot be predicted. For 90% of the time, many other events in your life can be forecast through processes such as trend analysis, scenario planning, DNA decoding, predictive analysis, intuition, human wisdom, and mental visualization of your preferred futures.

Q. Can your future be altered?

Yes it can. Making informed decisions today can create tangible benefits up ahead. You shape your future when you set goals, and learn fresh skills, and adjust to changing circumstances. When you seize opportunities for growth, your future life becomes a choice, rather than a matter of pre-determined fate.

Q. How might prediction help you to live a more fulfilling life?

The benefits are limitless. You can enhance your physical well-being, your mental health, your financial state. Anticipating future challenges can guide your choices and help you to be more prepared. You can contribute to your work and your community in many optimistic ways. You can raise your children to see the future as a promise, rather than a threat.

References

Chapter 1: Learning from the Past

1. *Magic and Superstition in the European Dark Ages*. History Is Now magazine. http://www.historyisnowmagazine.com/blog/2018/12/23/ magic-and-superstition-in-the-european-dark-ages#.ZE2l-HZBx2w

2. Kansas State University, *Psychology Professor Says Superstition All About Trying to Control Fate*. 28 October 2013. https://www. newswise.com/articles/psychology-professor-says-superstitions-all-about-trying-to-control-fate

3. Merriam-Webster. *Divining The Future*. https://www.merriam-webster.com/words-at-play/ways-to-tell-the-future

4. Bess Lovejoy. *10 Historical Divination Methods for Predicting the Future*. Mental Floss, 12 June 2019. https://www.mentalfloss.com/ article/585258/historical-divination-methods-predict-future

5. *Ten Strange Things You Didn't Know About the History of Magic*. Google Arts and Culture. https://artsandculture.google.com/story/ ten-strange-things-you-didn%E2%80%99t-know-about-the-history-of-magic/4ALSqk8PhNZ6KA

6. *Cartomancy*, https://en.wikipedia.org/wiki/Cartomancy Accessed on 2 February 2024.

7. *Tarot*, https://en.wikipedia.org/wiki/Tarot Accessed on 2 February 2024.

8. https://www.tarot.com/tarot/cards

9. *Palmistry*. Wikipedia. https://en.wikipedia.org/wiki/Palmistry Accessed on 2 February 2024.

10. *Crystal ball*. Wikipedia. https://en.wikipedia.org/wiki/Crystal_ball Accessed on 2 February 2024.

11. *History of magic*. Wikipedia. https://en.wikipedia.org/wiki/History_ of_magic Accessed on 2 February 2024.

12. Laura Hayward, *Oracle of Delphi: Why Was It So Important to Ancient Greeks?* 29 Nov 2020. https://www.thecollector.com/oracle-of-delphi/

13. *Divination: An Overview.* Encyclopedia.com, https://www.encyclopedia.com/environment/encyclopedias-almanacs-transcripts-and-maps/divination-overview

14. *Prediction.* Wikipedia. https://en.wikipedia.org/wiki/Prediction Accessed on 5 February 2024.

15. *How many people actually believe in astrology?* The Conversation, 28 April 2017. https://theconversation.com/how-many-people-actually-believe-in-astrology-71192

16. *What did Einstein say about astrology?* Quora. https://www.quora.com/What-did-Einstein-say-about-astrology

17. *Good Therapy*, Dream Analysis, 2 Feb 2016. https://www.goodtherapy.org/learn-about-therapy/types/dream-analysis

18. Stuart Jeffries, *War in Ukraine, death of the Queen, Elon Musk ... why are Nostradamus's 'predictions' still winning converts?* The Guardian, 10 October 2022. https://www.theguardian.com/books/2022/oct/10/why-nostradamus-predictions-are-still-winning-converts

19. Amy Sinclair, *Eerie Nostradamus prediction about Queen death stuns internet as major new royal upset forecast.* 7News, 12 September 2022. https://7news.com.au/lifestyle/eerie-nostradamus-prediction-about-queen-death-stuns-internet-as-major-new-royal-upset-forecast-c-8207977

20. *Baba Vanga.* Wikipedia. https://en.wikipedia.org/wiki/Baba_Vanga Accessed on 10 February 2024.

21. *Cassandra In Greek Mythology.* Greek Legends and Myths. https://www.greeklegendsandmyths.com/cassandra.html

Chapter 2: Understanding Prediction

1. Cheryl Critchley, *Every Day, We're Actually Seeing Into The Future.* 14 August 2018. The University of Melbourne. https://pursuit.unimelb.edu.au/articles/every-day-we-re-actually-seeing-into-the-future

2. Peter Sweeney. *Why Prediction is the Essence of Intelligence.* The Explainable Startup, June 2017. https://www.explainablestartup.com/2017/06/why-prediction-is-the-essence-of-intelligence.html

3. Matthew Syed. *Rebel Ideas: The Power of Thinking Differently.* London: John Murray, 2019.

4. Phillip Tetlock and Dan Gardner. *Superforecasting: The Art and Science of Prediction.* London: Random House, 2016.

5. *Prediction.* Wikipedia. https://en.wikipedia.org/wiki/Prediction Accessed on 27 January 2024.

6. Nassim Nicholas Taleb. *The Black Swan: The Impact of The Highly Improbable.* London: Penguin Books, 2010.

7. *Black Swans, Grey Swans, and White Swans.* Minet, 25 July 2022. http://www.minet.com/wp-content/uploads/2022/06/6.-Thought-leadership-Black-swans-Grey-swans-and-White-swans.pdf

8. *Popular Mechanics*, P.476. October 1909. https://books.google.com.au/books?id=nN8DAAAAMBAJ&lpg=PA476&vq=tesla&pg=PA476&redir_esc=y&hl=en#v=onepage&q&f=false

9. Luke Harding. *Numbers nerd Nate Silver's forecasts prove all right on election night.* The Guardian, 8 November 2012. https://www.theguardian.com/world/2012/nov/07/nate-silver-election-forecasts-right

10. https://fivethirtyeight.com/

Chapter 3: The Secrets of Probability

1. *Lottery Systems in Australia.* Lotto Simulation.https://lottosimulation.com/au/system-pick-entries

2. Robert Sanders. *Hindcasting helps scientists improve forecasts for life on Earth.* 12 June 2012, Berkeley News. https://news.berkeley.edu/2012/06/12/hindcasting-helps-scientists-improve-forecasts-for-life-on-earth/

3. Adam Zewe, MIT's New AI Model Predicts Human Behavior With Uncanny Accuracy. SciTech Daily, MIT, 19 April 2024. https://scitechdaily.com/mits-new-ai-model-predicts-human-behavior-with-uncanny-accuracy/

4. Tony Ryan, *The Next Generation: Preparing today's kids for an extraordinary future*. John Wiley & Sons Australia, 2017.

5. Adam Hayes. *YouTube Stats: Everything You Need to Know In 2023*. 23 February 2023, wyzowl. https://www.wyzowl.com/youtube-stats/

6. Maryann Mohsin. *10 Google Search Statistics You Need to Know In 2023*. 13 January 2023, Oberlo. https://www.oberlo.com/blog/google-search-statistics

7. Larry Dignan, *Are you suffering from filter failure?* Zdnet, 18 Sept 2008. https://www.zdnet.com/article/are-you-suffering-from-information-filter-failure/

8. James Purtill, *Prompt Engineers share tips on using ChatGPT, Midjourney, Bing Chat and other generative AI*. ABC News, 2 April 2023. https://www.abc.net.au/news/science/2023-04-02/prompt-engineers-share-their-tips-on-using-chatgpt-generative-ai/102165132

9. *Bayes' theorem*. Wikipedia. https://en.wikipedia.org/wiki/Bayes%27_theorem#Examples Accessed on 18 February 2024.

10. Bayes' Theorem Calculator. GIGA calculator. https://www.gigacalculator.com/calculators/bayes-theorem-calculator.php

11. Adam Hayes, *Bayes' Theorem: What It Is, the Formula, and Examples*. Investopedia, 1 March 2022. https://www.investopedia.com/terms/b/bayes-theorem.asp

12. Kavya Nambiar, *What is Laplace's Demon? Does this Demon Know Everything?* Science ABC, 15 Feb 2022. https://www.scienceabc.com/pure-sciences/what-is-laplaces-demon-definition-explanation.html

13. Adam Hayes. Game Theory. Investopedia, 28 March 2023. https://www.investopedia.com/terms/g/gametheory.asp

14. *Karen Willcox Lays Out Blueprint for Future of Digital Twins at TEDxUTAustin*. Aerospace Engineering and Engineering Mechanics, 24 March 2022. https://www.ae.utexas.edu/news/karen-willcox-lays-out-blueprint-for-future-of-digital-twins-at-tedxutaustin

15. Edwin O. *This Invention has created the first Digital Twin in history: a copy of you, in human form and for this purpose.* EcoNews, 29 August 2024. https://www.ecoticias.com/en/digital-twin-ai-intervention/5800/

16. Savannah Sher. *Top 10 Funniest Tik Tok Trends.* watch mojo. https://www.watchmojo.com/articles/top-10-funniest-tiktok-trends

17. https://www.apf.org/compass

18. *Superforecaster*, Wikipedia. https://en.wikipedia.org/wiki/Superforecaster Accessed on 27 February 2023.

19. https://goodjudgment.com/

20. *Wisdom of the crowd.* Wikipedia. https://en.wikipedia.org/wiki/Wisdom_of_the_crowd Accessed on 15 July 2024.

21. Andy Owen, Superfore*casters in the Cosmic Bazaar. https://www.arcdigital.media/p/superforecasters-in-the-cosmic-bazaar*

22. Phillip Tetlock and Dan Gardner. *Superforecasting: The Art and Science of Prediction.* London: Random House, 2016.

Chapter 4: Practical Prediction

1. Jason Guerrasio, *21 Times 'The Simpsons' accurately predicted the future.* Insider, 20 Aug 2022. https://www.insider.com/the-simpsons-is-good-at-predicting-the-future-2016-11

2. *Up (film series).* Wikipedia. https://en.wikipedia.org/wiki/Up_(film_series) Accessed on 12 April 2024.

3. https://dunedinstudy.otago.ac.nz/studies

4. Tony Ryan, *The Next Generation: Preparing today's kids for an extraordinary future.* John Wiley & Sons Australia, 2017.

5. Joe Satran, *'Your Cell Phone Could Soon Predict Whether You'll Get the Flu',* The Huffington Post, 21 August 2015. www.huffingtonpost.com.au/entry/cell-phone-predict-flu_us_55d4c780e4b0ab468d9f6294

6. *Making Predictions.* Reading Rockets. https://www.readingrockets.org/article/making-predictions

7. Thi Thibeaux. *6 Sci-Fi Movies That Actually Predicted the Future.* Movieweb, 19 December 2022. https://movieweb.com/sci-fi-movies-predicted-the-future/#the-running-man

8. *Crystal Ball Gazing: Books that Predicted the Future.* Public Libraries Singapore, 22 January 2021. https://medium.com/publiclibrarysg/crystal-ball-gazing-books-that-predicted-the-future-968469d9fc89

9. Future Shock: *11 Real-Life Technologies That Science Fiction Predicted.* Micron, 2023. https://www.micron.com/insight/future-shock-11-real-life-technologies-that-science-fiction-predicted

10. *Roger Federer may be able to predict Rafael Nadal's moves, thanks to new algorithm.* The Economic Times, 23 January 2019. https://economictimes.indiatimes.com/magazines/panache/roger-federer-may-be-able-to-predict-rafael-nadals-moves-thanks-to-new-algorithm/articleshow/67657957.cms

11. Drew Hutchins. *How to Read a Soccer Penalty Shot if You're a Goalie.* 6 December 2022, WikiHow. https://www.wikihow.com/Read-a-Soccer-Penalty-Shot-if-You%27re-a-Goalie

12. *How many moves do chess players think ahead?* Rookie Road. https://www.rookieroad.com/chess/how-many-moves-do-chess-players-think-ahead-2941395/

13. Jean Folger, *How to Calculate Insurance Premiums.* Investopedia, 18 March 2022. https://www.investopedia.com/ask/answers/09/calculating-premium.asp

14. Mary Hall. What Affects Profit Margins in the Insurance Sector? Investopedia, 30 August 2023. https://www.investopedia.com/ask/answers/052515/what-usual-profit-margin-company-insurance-sector.asp

15. Mark Rosanes. These are the world's 20 largest insurance companies in 2022. Insurance Business Mag, 23 September 2022. https://www.insurancebusinessmag.com/us/guides/these-are-the-worlds-20-largest-insurance-companies-in-2022-421548.aspx

16. Genetic Testing for Cancer Risk. Cancer.Net, September 2018. https://www.cancer.net/navigating-cancer-care/cancer-basics/genetics/genetic-testing-cancer-risk

17. Karen Weintraub. *Want to know when you're going to die? Your lifespan is written in your DNA, and we're learning to read the code.* MIT Technology Review, 19 October 2018. https://www. technologyreview.com/2018/10/19/139463/want-to-know-when-youre-going-to-die/

18. David Cox, *How genetics determine our life choices.* BBC Future, 11 May 2023. https://www.bbc.com/future/article/20230509-how-genetics-determine-our-life-choices

19. Morela Hernandez, *How Previous Generations Influence Our Decisions.* MIT Sloan, 1 July 2019. https://sloanreview.mit.edu/article/how-previous-generations-influence-our-decisions/

20. *How Intergenerational Trauma Impacts Families.* Psych Central. https://psychcentral.com/lib/how-intergenerational-trauma-impacts-families

21. Jennifer Sass, *Scientific Evidence to Support 'Seven Generations' future thinking; our toxic chemical exposures may harm our great-grandchildren.* NRDC, 28 November 2013. https://www.nrdc.org/bio/jennifer-sass/scientific-evidence-support-seven-generations-future-thinking-our-toxic-chemical

22. Michael Gottschalk and Katharina Domschke. *Dialogues in Clinical Neuroscience. Genetics of generalized anxiety disorder and related traits.* National Library of Medicine, June 2017. https://www.ncbi.nlm. nih.gov/pmc/articles/PMC5573560/

23. Kerry McGee. *Are Anxiety Disorders Genetic?* GoodRx Health, 16 March 2023. https://www.goodrx.com/health-topic/anxiety-disorders/is-anxiety-genetic-or-hereditary

24. *Sliding Doors.* Wikipedia. https://en.wikipedia.org/wiki/Sliding_ Doors Accessed on 23 July 2024.

25. *World War 1.* 13 July 2023, History. https://www.history.com/topics/world-war-i/world-war-i-history

26. Corinne McLaughlin & Gordon Davidson. *Spiritual Politics: Changing the World from The Inside-Out.* New York: Ballantine Books, 1994.

27. Tony Ryan. *The Ripple Effect: How You Can Make a Difference to the World Every Day.* Australia: Headfirst Publishing, 2000.

Chapter 5: Prediction By Mother Nature

1. *Mission San Juan Capistrano.* Wikipedia. https://en.wikipedia.org/wiki/Mission_San_Juan_Capistrano#Return_of_the_swallows Accessed on 2 April 2024.

2. *Arctic tern.* Wikipedia. https://en.wikipedia.org/wiki/Arctic_tern Accessed on 5 April 2024.

3. *Human Behavior is 93% Predictable, Research Shows.* Northwestern University, 19 February 2010. https://cos.northeastern.edu/news/human-behavior-is-93-predictable-research-shows/

4. Mark Buchanan, *Why we are all creatures of habit.* 4 July 2007, New Scientist.

5. Gerardo Sison, *Does Your Body Really Replace Itself Every 7 Years?* Discovery, 1 August 2019. https://www.discovery.com/science/Body-Really-Replace-Itself-Every-7-Years

6. *7 amazing Facts About Periods That Everyone Needs To Know.* 9 April 2019, helping women period. https://www.helpingwomenperiod.org/7-amazing-facts-about-periods-that-everyone-needs-to-know/

7. Mayo Clinic Staff. *Male Menopause: Myth or reality?* 24 May 2022, Mayo Clinic. https://www.mayoclinic.org/healthy-lifestyle/mens-health/in-depth/male-menopause/art-20048056

8. Aaron Robotham and Sabine Bellstedt. *Humans have been predicting eclipses for thousands of years, but it's harder than you think.* 17 April 2023, The University of Western Australia. https://www.uwa.edu.au/news/Article/2023/April/Humans-have-been-predicting-eclipses-for-thousands-of-years-but-its-harder-than-you-might-think

9. Robert Lea. *This is what would happen if scientists found an asteroid heading to Earth.* 6 April 2023, space.com. https://www.space.com/nasa-models-hypothetical-asteroid-impact-scenario

10. *Solar Cycle*, Wikipedia. https://en.wikipedia.org/wiki/Solar_cycle Accessed on 6 April 2024.

11. *Early Weather Observations.* NOAA. https://celebrating200years.noaa.gov/foundations/weather_obs/side_early_obs.html

12. *Ring Around the Moon? Here's What It Means.* Farmers' Almanac, 17 April 2023. https://www.farmersalmanac.com/ring-around-the-moon-9657

13. *Using Clouds To Predict The Weather.* The Home School Scientist, 2023. https://thehomeschoolscientist.com/using-clouds-to-predict-the-weather/

14. *Weather Forecasting Through the Ages.* NASA Earth Observatory, 25 Feb 2002. https://earthobservatory.nasa.gov/features/WxForecasting/wx2.php

15. Todd Dankers, *AI and machine learning are improving weather forecasts, but they won't replace human experts.* The Conversation, 26 May 2022. https://theconversation.com/ai-and-machine-learning-are-improving-weather-forecasts-but-they-wont-replace-human-experts-182498

16. Erik Nymann. *12 Animals That Can Supposedly Predict the Weather.* Weather Station Advisor, 20 July 2022. https://www.weatherstationadvisor.com/animals-that-can-predict-weather/

17. *Can Animals Predict the Weather?* 25StormGeo. https://stormgeo.com/insights/can-animals-predict-the-weather

18. Sandi Schwartz, *12 Ways to Predict the Weather by Watching Nature in Your Backyard.* Bob Vila, 18 Oct 2022. https://www.bobvila.com/articles/predict-the-weather-by-watching-nature/

19. *UN's global disaster alert systems goal faces uphill battle.* news.com.au, 24 March 2023. https://www.news.com.au/breaking-news/uns-global-disaster-alert-systems-goal-faces-uphill-climb/news-story/562e698c350e27e3426b03680830d917

20. Alok Jha, *Toads able to detect earthquake days beforehand, says study.* The Guardian, 31 May 2010. https://amp.theguardian.com/science/2010/mar/31/toads-detect-earthquakes-study

21. *Can animals predict earthquakes?* USGS. https://www.usgs.gov/faqs/can-animals-predict-earthquakes

22. *How Android Earthquake Systems Work.* Google Crisis Response. https://crisisresponse.google/android-alerts/

23. *Arielle Duhaime*-Ross, Manslaughter conviction overturned for Italian geologists, but other scientists are still fearful. 12 November 2014. The Verge. https://www.theverge.com/2014/11/11/7193391/italy-judges-clear-geologists-manslaughter-laquila-earthquake-fear

24. John Cook (ed.), *Global Warming and Climate Change Myths*. Skeptical Science, 2023. https://skepticalscience.com/argument.php

25. IPCC, *Climate change widespread, rapid, and intensifying*. IPCC, 9 Aug 2021. https://www.ipcc.ch/2021/08/09/ar6-wg1-20210809-pr/

26. List of weather records. Wikipedia. *https://en.wikipedia.org/wiki/List_of_weather_records* Accessed on 16 July 2024.

27. Andrew Nikiforuk, *The Rising Chorus of Renewable Energy Skeptics*. Post Carbon Institute, 10 April 2023. https://www.resilience.org/stories/2023-04-10/the-rising-chorus-of-renewable-energy-skeptics/

28. *The world's biggest ecosystem restoration project*. UN Environment Program, 23 April 2020. https://www.unep.org/news-and-stories/story/worlds-biggest-ecosystem-restoration-project

29. https://theoceancleanup.com/

30. https://futurecrunch.com/

31. https://fixthenews.com/

32. https://goodnews.eu/en/

33. James Atkinson, *Get to know the five Rs*. Clean Up Australia. https://www.cleanup.org.au/the5rs

34. *Climate engineering*. Wikipedia. https://en.wikipedia.org/wiki/Climate_engineering Accessed on 17 July 2024.

35. Elizabeth Gamillo, *This Metal-Rich, Potato-Shaped Asteroid Could Be Worth $10 Quintillion*. Smithsonian Magazine, 4 January 2022. https://www.smithsonianmag.com/smart-news/asteroid-16-psyche-may-be-worth-more-than-planet-earth-at-10-quintillion-in-fine-metals-180979303/

36. 700+ cities in 53 countries now committed to halve emissions by 2030 and reach net zero by 2050. C40 Cities, 16 April 2021. https://www.c40.org/news/cities-committed-race-to-zero/

37. Sean Fleming, *What makes Copenhagen the most bike-friendly city?* World Economic Forum, 5 Oct 2018. https://www.weforum.org/ agenda/2018/10/what-makes-copenhagen-the-worlds-most-bike-friendly-city/

Chapter 6: Life's A Gamble

1. *Paul the Octopus.* Wikipedia. https://en.wikipedia.org/wiki/Paul_the_ Octopus Accessed on 25 April 2024.

2. *Jack Andrews.* How To Be A Professional Gambler - Mentality. 20 January 2023. Unabated. https://unabated.com/articles/professional-sports-gambler-mentality

3. *The world's biggest gamblers.* 9 February 2017, The Economist. https://www.economist.com/graphic-detail/2017/02/09/the-worlds-biggest-gamblers

4. M. Browne, N. Greer, T. Armstrong, I. Kinchin, E. Langham, M. Rockloff. *The social cost of gambling to Victoria.* Victorian Responsible Gambling Foundation, November 2017. https://responsiblegambling. vic.gov.au/documents/121/research-social-cost-of-gambling.pdf

5. Dawn Teh, *The psychology and brain science of a gambling addiction.* Health Match, 19 May 2022. https://healthmatch.io/blog/the-psychology-and-brain-science-of-a-gambling-addiction

6. J. B. Maverick, *Why Does the House Always Win? A Look at Casino Profitability.* Investopedia, 29 November 2022. https://www. investopedia.com/articles/personal-finance/110415/why-does-house-always-win-look-casino-profitability.asp

7. Maggie Harrison, *A Math and Physics Savant Reportedly Figured Out How to Beat Roulette.* The Byte. https://futurism.com/the-byte/math-physics-savant-beat-roulette

8. *Numerology.* Wikipedia. Accessed on 17 July 2024. https:// en.wikipedia.org/wiki/Numerology

9. *Lucky and Unlucky Numbers Around the World.* Cultural Mixology. https://culturalmixology.com/lucky-and-unlucky-numbers-around-the-world/

10. What is the Impact of Numerology on your Life. Astrobix. https://astrobix.com/numero/45-what-is-the-impact-of-numerology-on-your-life.html

11. *Lottery Systems in Australia.* Lotto Simulation. https://lottosimulation.com/au/system-pick-entries

12. *How retired couple outsmarted the lotto system to win millions.* news.com.au, 15 August 2022. https://www.news.com.au/finance/business/media/how-retired-couple-outsmarted-the-lotto-system-to-win-millions/news-story/883707997919aa693cc889a859f28460

13. Gary Drevitch. *Seven Reasons We Are Captivated by the Number Seven.* 27 June 2015, Psychology Today. https://www.psychologytoday.com/au/blog/the-squeaky-wheel/201506/seven-reasons-we-are-captivated-the-number-seven

14. Dharmesh Tolia. *The Importance of Number 7.* 13 December 2018, linkedin. https://www.linkedin.com/pulse/importance-number-7-dharmesh-tolia-l-i-o-n-/

15. Mo Abdelbaki, *The Seven Stages of Life, and Then Some.* Gaia, 9 Feb 2020. https://www.gaia.com/article/seven-cycles-of-life

16. *Sports Prediction: Creating a Sports Betting Model 101.* Primo.ai. http://primo.ai/index.php?title=Sports_Prediction

17. Barry D. Moore, *Elliott Wave Theory: Examples and Reliability Uncovered.* Liberated Stock Trader, 16 May 2023. https://www.liberatedstocktrader.com/elliott-wave-theory-principle-examples-stock-market/

18. Mary Hall, *Market Cycles: The Key to Maximum Returns.* Investopedia, 19 July 2021. https://www.investopedia.com/trading/market-cycles-key-maximum-returns/

19. Adam Hayes. *Market Psychology: What is it, Predictions, and FAQ.* 11 May 2022, Investopedia. https://www.investopedia.com/terms/m/marketpsychology.asp

20. *History of bitcoin.* Wikipedia. https://en.wikipedia.org/wiki/History_of_bitcoin Accessed on 27 April 2024.

21. Nasdaq finds scams led to $486 billion in losses in 2023. ABA Banking Journal, 19 January 2024. https://bankingjournal.aba.com/2024/01/nasdaq-finds-scams-led-to-486-billion-in-losses-in-2023/

22. Danny Tran. *Melbourne racing identity Bill Vlahos jailed for nine years for $17.5m punting Ponzi punting scheme.* ABC News, 17 Dec 2021. https://www.abc.net.au/news/2021-12-17/bill-vlahos-conman-racing-ponzi-scheme-melbourne-betting/100708368

Chapter 7: Intuitive Prediction

1. *Psychic Scams.* AARP, 28 April 2022. https://www.aarp.org/money/scams-fraud/info-2022/psychic.html

2. Miriam Frankel, *Great Mysteries of Physics: is time an illusion?* The Conversation, 8 March, 2023. https://theconversation.com/great-mysteries-of-physics-1-is-time-an-illusion-201026

3. *Quantum mechanics.* Wikipedia. https://en.wikipedia.org/wiki/Quantum_mechanics Accessed on 17 July 2024.

4. Alan Boyle. *Are quantum computers for real? So far, the uncertainty principle rules the day.* 4 February 2023. GeekWire. https://www.geekwire.com/2023/quantum-computers-uncertainty-principle/

5. Kim Arlington, *Supernatural sleuths and the search for truth.* SMH, 30 December 2010. https://www.smh.com.au/national/supernatural-sleuths-and-the-search-for-truth-20101229-19a9z.html

6. Jeremy Sutton, *What Is Intuition and Why Is It Important? Five Examples.* PositivePsychology.com, 27 August 2020. https://positivepsychology.com/intuition/

7. Daniel Kahneman, *Thinking Fast And Slow.* New York. Farrar, Straus and Giroux, 2011.

8. *A Guide to Developing Intuition: 12 Ways to Tap into Your Inner Knowing.* SkillShare Blog, 27 May 2021. https://www.skillshare.com/en/blog/a-guide-to-developing-intuition-12-ways-to-tap-into-your-inner-knowing/

9. Michael Shermer, *Rupert's Resonance*. Scientific American, 1 November 2005. https://www.scientificamerican.com/article/ruperts-resonance/

10. John Horgan, *Sheldrake on Morphic Fields, Psychic Dogs, and Other Mysteries*. Scientific American, 14 July 2014. https://blogs.scientificamerican.com/cross-check/scientific-heretic-rupert-sheldrake-on-morphic-fields-psychic-dogs-and-other-mysteries/

11. Venkat S.R., *What Is an Empath?* WebMD, 9 November 2022. https://www.webmd.com/balance/what-is-an-empath

12. Judith Orloff, *4 Signs You Might Be an Empath*. Oprah.com. https://www.oprah.com/inspiration/judith-orloff-md-4-signs-you-might-be-an-intuitive-empath

13. Giacomo Rizzolatti and Laila Graighero, *Mirror neuron: a neurological approach to empathy*. Springer Link. https://link.springer.com/chapter/10.1007/3-540-29803-7_9

14. *Can we time travel? A theoretical physicist provides some answers*. The Conversation. 14 June 2022. https://theconversation.com/can-we-time-travel-a-theoretical-physicist-provides-some-answers-182634

15. Callum Hoare. *Time travel Proof? How Brian Cox Claimed Future Travel Is Possible*. Express, 12 April 2019. https://www.express.co.uk/news/science/1112960/time-travel-news-proof-brian-cox-claims-travel-future-albert-einstein-physics-spt

16. Alyssa Danigelis. *Machine Being Built to Receive Messages from the Future*. Seeker, 3 Nov 2015. https://www.seeker.com/machine-being-built-to-receive-messages-from-the-future-1770421978.html

Chapter 8: Exponential Futures

1. *Moore's Law*. Wikipedia. https://en.wikipedia.org/wiki/Moore%27s_law Accessed on 15 July 2024.

2. William D. Sheridan, *Exponential Change Applies to Everything – Not Just Technology*. Business Learning Institute, 4 January 2021. https://blionline.org/news/2172-exponential-change-applies-to-everything-not-just-technology-2021-01-04

3. *Clarke's three laws*. New Scientist. https://www.newscientist.com/definition/clarkes-three-laws/

4. *Transhumanism: Savior of humanity or false prophecy?* Big Think, 27 July 2022. https://bigthink.com/the-future/transhumanism-savior-humanity-false-prophecy/

5. *Technological singularity*. Wikipedia. https://en.wikipedia.org/wiki/Technological_singularity Accessed on 22 July 2024.

6. Tony Ryan, *The Next Generation: Preparing today's kids for an extraordinary future*. John Wiley & Sons Australia, 2017.

7. *Global Health Estimates: Life expectancy and healthy life expectancy*. World Health Organisation. https://www.who.int/data/gho/data/themes/mortality-and-global-health-estimates/ghe-life-expectancy-and-healthy-life-expectancy

8. *Living Too Long*. Embo Reports, 18 December 2014. https://www.ncbi.nlm.nih.gov/pmc/articles/PMC4328740/

9. Dan Tynan, *Augmented Eternity: scientists aim to let us speak from beyond the grave*, The Guardian, 23 June 2016. https://www.theguardian.com/technology/2016/jun/23/artificial-intelligence-digital-immortality-mit-ryerson

10. Verity Linehan, Sally Thorpe, Neil Andrews, Yeon Kim and Farah Beaini, *Food demand to 2050*. Department of Agriculture, Fisheries and Forestry, Australian Government, 4 March 2012. https://www.agriculture.gov.au/sites/default/files/sitecollectiondocuments/abares/publications/Outlook2012FoodDemand2050.pdf

11. Stephanie Safdie, *Global Food Waste in 2023*. Greenly Resources, 24 March 2023. https://greenly.earth/en-us/blog/ecology-news/global-food-waste-in-2022

12. George Monbiot. *With our food systems on the verge of collapse, it's the plutocrats v life on Earth*. 15 July 2023, The Guardian. https://www.theguardian.com/commentisfree/2023/jul/15/food-systems-collapse-plutocrats-life-on-earth-climate-breakdown

13. *How Climate Change Impacts Water Access*. National Geographic. https://education.nationalgeographic.org/resource/how-climate-change-impacts-water-access/

14. *18 Surprising Projections About the Future of Water*. Seametrics. https://www.seametrics.com/blog/future-water/

15. *Desalinating seawater sounds easy, but there are cheaper and more sustainable ways to meet people's water needs*. The Conversation, 22 Sept 2022. https://theconversation.com/desalinating-seawater-sounds-easy-but-there-are-cheaper-and-more-sustainable-ways-to-meet-peoples-water-needs-184919

16. Mark Wilkinson, *The 50 Weirdest Job Titles*. https://coburgbanks.co.uk/blog/friday-funnies/the-50-weirdest-job-titles/

17. Tytler, R., Bridgstock, R., White, P., Mather, D., McCandless, T., Grant-Iramu, M., Bonson, S., Ramnarine, D., & Penticoss, A.J. *100 Jobs of The Future*. Ford Australia, Deakin University, Griffith University, 23 July 2019. https://100jobsofthefuture.com/browse/

18. Colin Milner, Ken Dychtwald: How the Age Wave pathfinder inspired a new view of aging. September 2018. ICAA. https://www.icaa.cc/blog/Ken%20Dychtwald-Visionaries.pdf

Chapter 9: Developing Your Own Predictive Future

1. Jory MacKay. *Affective forecasting: Why you keep giving "future you" too much work*. RescueTime, 20 November 2018. https://blog.rescuetime.com/affective-forecasting/

2. Nicole Celestine, *What is Affective Forecasting? A Psychologist Explains*. 1 Sept 2018 reviewed 7 April 2023. https://positivepsychology.com/affective-forecasting/

3. Kendra Cherry. What Is Cognitive Bias? 7 November 2022, verywell mind. https://www.verywellmind.com/what-is-a-cognitive-bias-2794963

4. Negativity bias. Wikipedia. https://en.wikipedia.org/wiki/Negativity_bias Accessed on 18 July 2024.

5. *Why is the news always so depressing? The Negativity Bias, explained*. The Decision Lab. https://thedecisionlab.com/biases/negativity-bias#

6. *Self-Fulfilling Prophecy.* ScienceDirect. https://www.sciencedirect. com/topics/psychology/self-fulfilling-prophecy

7. Michelle P. Maidenberg. *Beware of Your Self-Fulfilling Prophecy.* Psychology Today, 12 October 2021. https://www.psychologytoday. com/us/blog/being-your-best-self/202110/beware-your-self-fulfilling-prophecy

8. University of Bath. *Women feel the pain of losses more than men when faced with risky choices, according to new research.* Medical Press, 9 June 2023. https://medicalxpress.com/news/2023-06-women-pain-losses-men-risky.html

9. Jennifer Smith. Growth Mindset vs Fixed Mindset: How what you think affects what you achieve. 25 September 2020, mindset health. https:// www.mindsethealth.com/matter/growth-vs-fixed-mindset

10. Dillon Jacobs. *5 Steps to Improving Your Prediction Skills.* Finmasters, 26 November 2022. https://finmasters.com/improve-prediction-skills/#gref

11. *Predictive Analytics: What it is and why it matters.* SAS. https://www. sas.com/en_ae/insights/analytics/predictive-analytics.html

12. Gregg Schwartz, *12 Tactics for Better Sales Forecasting + 5 Forecasting Models to Leverage.* HubSpot, 15 November 2021. https:// blog.hubspot.com/sales/accurate-sales-forecasting-model-tips

13. Alexandra Twin, *Delphi Method Forecasting: Definition and How It's Used.* 27 May 2022. Investopedia. https://www.investopedia.com/ terms/d/delphi-method.asp

14. David Luther and Rami Ali, Scenario Planning: Strategy, Steps and Practical Examples. 25 August 2022, Netsuite. https://www.netsuite. com/portal/resource/articles/financial-management/scenario-planning.shtml

15. *The Scientific Method.* Khan Academy. https://www.khanacademy.org/ science/biology/intro-to-biology/science-of-biology/a/the-science-of-biology

Chapter 10: Developing The Planet's Predictive Future

1. Charlotte Akers, *Seven Generation Thinking*. EcoResolution. https://www.ecoresolution.earth/resources/seven-generation-thinking

2. Richard Fisher, The perils of short-termism: Civilization's greatest threat. 10 January 2019, BBC Future. https://www.bbc.com/future/article/20190109-the-perils-of-short-termism-civilisations-greatest-threat

3. *Bulletin of the Atomic Scientists*. https://thebulletin.org/

4. Tony Ryan, *The Next Generation: Preparing today's kids for an extraordinary future*. John Wiley & Sons Australia, 2017.

5. *List of dates predicted for apocalyptic events*. Wikipedia..https://en.wikipedia.org/wiki/List_of_dates_predicted_for_apocalyptic_events Accessed on 8 June 2024.

6. Reza Aslan, *How religions predict the world will end*. cnn.com, 10 March 2017. https://edition.cnn.com/2017/03/09/world/gallery/believer-end-of-world-prophecies/index.html

7. Edward Helmore, *Yes, it's bleak, says expert who tested 1970s end-of-the-world prediction*. The Guardian, 25 July 2021. https://www.theguardian.com/environment/2021/jul/25/gaya-herrington-mit-study-the-limits-to-growth

8. *Societal collapse*. Wikipedia. https://en.wikipedia.org/wiki/Societal_Collapse Accessed on 8 June 2024.

9. Mark G. Edwards, *The growth paradox, sustainable development, and business strategy*. Wiley Online Library, 4 May 2021. https://onlinelibrary.wiley.com/doi/full/10.1002/bse.2790

10. Gregg Easterbrook. *The Progress Paradox: How life gets better while people feel worse*. New York, Random House, 2003.

11. Tony Ryan, *The Next Generation: Preparing today's kids for an extraordinary future*. John Wiley & Sons Australia, 2017.

12. Larry Elliott, *Each generation should be better off than their parents? Think again.* The Guardian, 14 February 2016. https://www. theguardian.com/business/2016/feb/14/economics-viewpoint-baby-boomers-generation-x-generation-rent-gig-economy

13. Max Roser, *The short history of global living conditions and why it matters that we know it.* Our World in Data, 2020. https:// ourworldindata.org/a-history-of-global-living-conditions

14. *What Worries The World?* Ipsos, December 2022. https://www.ipsos. com/sites/default/files/ct/news/documents/2023-01/Global%20 Report%20-%20What%20Worries%20the%20World%20Dec22.pdf

15. Jacob Poushter, Moira Fagan, Sneha Gubbala. *Climate Change Remains Top Global Threat Across 19-Country Survey.* Pew Research Center, 31 August 2022. https://www.pewresearch.org/ global/2022/08/31/climate-change-remains-top-global-threat-across-19-country-survey/

16. Roger Harrabin, Climate change: *Young people very worried - survey.* BBC News, 14 September 2021. https://www.bbc.com/news/ world-58549373

17. Jessica Leggett & Natalie Wolchover, *20 inventions that changed the world.* LiveScience, 14 February 2023.. https://www.livescience. com/33749-top-10-inventions-changed-world.html

18. Rutger Bregman. *Humankind: A Hopeful History.* London: Bloomsbury Publishing PLC, 2020. https://www.xprize.org/

Reference Acknowledgement to Chat-GPT:

All material in every chapter was written exclusively by the author. Chat-GPT was used in part for the 'Preface', the 'Predictive Exercises' and the 'In Summary' sections at the end of each chapter. In all cases, those parts were based upon initial written material from the author.

Bibliography

Catherine Ball. *Converge: A futurist's insight into the potential of our world as technology and humanity collide.* Melbourne: Major Street Publishing, 2022.

Rutger Bregman. *Humankind: A Hopeful History.* London: Bloomsbury Publishing PLC, 2020.

James Clear. *Atomic habits: tiny changes, remarkable results: an easy & proven way to build good habits & break bad ones.* New York: Avery, an imprint of Penguin Random House, 2018.

Geoffrey Colvin. *Humans Are Underrated: What High Achievers Know That Brilliant Machines Never Will.* New York: Portfolio Penguin, 2015.

Peter Diamandis. *Abundance: The Future Is Better Than You Think.* Massachusetts: Free Press, 2012.

Peter Diamandis and Steven Kotler. *Bold: How to Go Big, Create Wealth, and Impact the World.* New York: Simon & Schuster, 2015.

Peter Diamandis and Steven Kotler. *The Future Is Faster Than You Think: How Converging Technologies Are Transforming Business, Industries, and Our Lives.* New York: Simon & Schuster, 2020.

Ken Dychtwald and Joe Flower. *Age Wave: How the Most Important Trend of Our Time Will Change Your Future.* Bantam Books: USA, 1990.

Tim Dunlop. *Why the Future Is Workless.* Sydney: New South Publishing, 2016.

Martin Ford. *The Rise of the Robots: Technology and the Threat of Mass Unemployment.* London: One World Publications, 2015.

Andrew Fuller. *The A to Z of Feelings.* Australia, Bad Apple Press, 2021.

Dan Gardner. Future Babble. *Why Expert Predictions Fail — And Why We Believe Them Anyway.* Australia: Scribe Publications, 2011.

Adam Grant. *Think Again: The Power of Knowing What You Don't Know.* Penguin Publishing Group, 2021

Daniel Goleman. *Focus: The Hidden Driver of Excellence*. London: Bloomsbury, 2013.

Gladwell, Malcolm. *Blink: The Power of Thinking Without Thinking*. New York: Little, Brown and Co., 2005.

Steve Hilton. *More Human: Designing a World Where People Come First*. London: WH Allen, 2015.

Steven Johnson. *Where Good Ideas Come From: The Natural History of Innovation*. London: Penguin, 2010.

Daniel Kahneman. *Thinking Fast and Slow*. New York: Farrar, Straus and Giroux, 2011.

Daniel Pink. *Drive: The Surprising Truth about What Motivates Us*. USA: Riverhead Books, 2009.

Steven Pinker. *The Better Angels of Our Nature: Why Violence Has Declined*. New York: Viking, 2011.

Tony Ryan. *Thinkers Keys: A Powerful Program for Teaching Children to Become Extraordinary Thinkers*. Brisbane: Headfirst Publishing, 2014.

Tony Ryan. *The Ripple Effect: How You Can Make a Difference to the World Every Day*. Brisbane: Headfirst Publishing, 1999.

Tony Ryan. *The Next Generation: Preparing today's kids for an extraordinary future*. John Wiley & Sons Australia, 2017.

Mustafa Suleyman with Michael Bhaskar. *The Coming Wave: AI, Power and the 21st Century's Greatest Dilemma*. UK: Penguin Random House, 2023.

Matthew Syed. *Rebel Ideas: The Power of Thinking Differently*. London: John Murray, 2019.

James Surowiecki. *The Wisdom of Crowds: Why the Many Are Smarter Than the Few and How Collective Wisdom Shapes Business, Economies, Societies and Nations*. Anchor, 2005.

Nassim Nicholas Taleb. *The Black Swan: The Impact of The Highly Improbable*. London: Penguin Books, 2010.

Phillip Tetlock and Dan Gardner. *Superforecasting: The Art and Science of Prediction*. London: Random House, 2016.